MW00415215

GEOLOGY
OF NEW BRUNSWICK AND PRINCE EDWARD ISLAND
Touring through time at 44 scenic sites

Martha Hickman Hild and Sandra M. Barr

Other field guides by Boulder Books

Geology of Newfoundland by Martha Hickman Hild

Geology of Nova Scotia by Martha Hickman Hild and Sandra M. Barr

Birds of Newfoundland by Ian Warkentin and Sandy Newton

Edible Plants of Atlantic Canada by Peter J. Scott

Edible Plants of Newfoundland and Labrador by Peter J. Scott

Trees & Shrubs of the Maritimes by Todd Boland

Trees & Shrubs of Newfoundland and Labrador by Todd Boland

Stouts, Millers, and Forky-Tails: Insects of Newfoundland
by Tom Chapman, Peggy Dixon, Carolyn Parsons, and Hugh Whitney

Wildflowers and Ferns of Newfoundland by Todd Boland

Wildflowers of Nova Scotia by Todd Boland

Wildflowers of New Brunswick by Todd Boland

Gardening guides by Boulder Books

Atlantic Gardening by Peter J. Scott

Newfoundland Gardening by Peter J. Scott

Favourite Perennials for Atlantic Canada by Todd Boland

GEOLOGY

OF NEW BRUNSWICK AND PRINCE EDWARD ISLAND

Touring through time at 44 scenic sites

Martha Hickman Hild and Sandra M. Barr

Library and Archives Canada Cataloguing in Publication

Title: Geology of New Brunswick and Prince Edward Island : field guide / Martha Hickman Hild & Sandra M. Barr.

Names: Hild, Martha Hickman, author. | Barr, Sandra M., author.
Description: Includes bibliographical references and index.
Identifiers: Canadiana 20190130687 | ISBN 9781999491031 (softcover)
Subjects: LCSH: Geology—New Brunswick—Guidebooks. | LCSH: Geology—Prince Edward Island—Guidebooks.
| LCSH: Historical geology—New Brunswick. | LCSH: Historical geology—Prince Edward Island. |
LCSH: New Brunswick—Guidebooks. | LCSH: Prince Edward Island—Guidebooks. | LCGFT: Guidebooks.
Classification: LCC QE185 .H55 2019 | DDC 557.15—dc23

Book concept: Martha Hickman Hild
Editor: Stephanie Porter
Copy editor: Iona Bulgin
Cover design and page layout: Tanya Montini

Front cover: Hopewell Rocks, New Brunswick (Kevin Snair)
Back cover: Cavendish Beach, Prince Edward Island

Printed in China

We acknowledge the financial support of the Government of Newfoundland and Labrador through the Department of Tourism, Culture and Recreation.

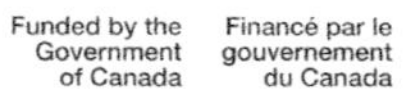

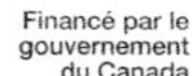

Dedication

To the memory of Sandra's parents, Vernon Wesley Barr (1921–2017) and Ruth Marie Barr, née Sawyer (1923–2018), of St. Stephen, New Brunswick.

They gave her the opportunity and freedom to explore the natural history of southern New Brunswick and Prince Edward Island, thus setting her on a lifelong geological journey.

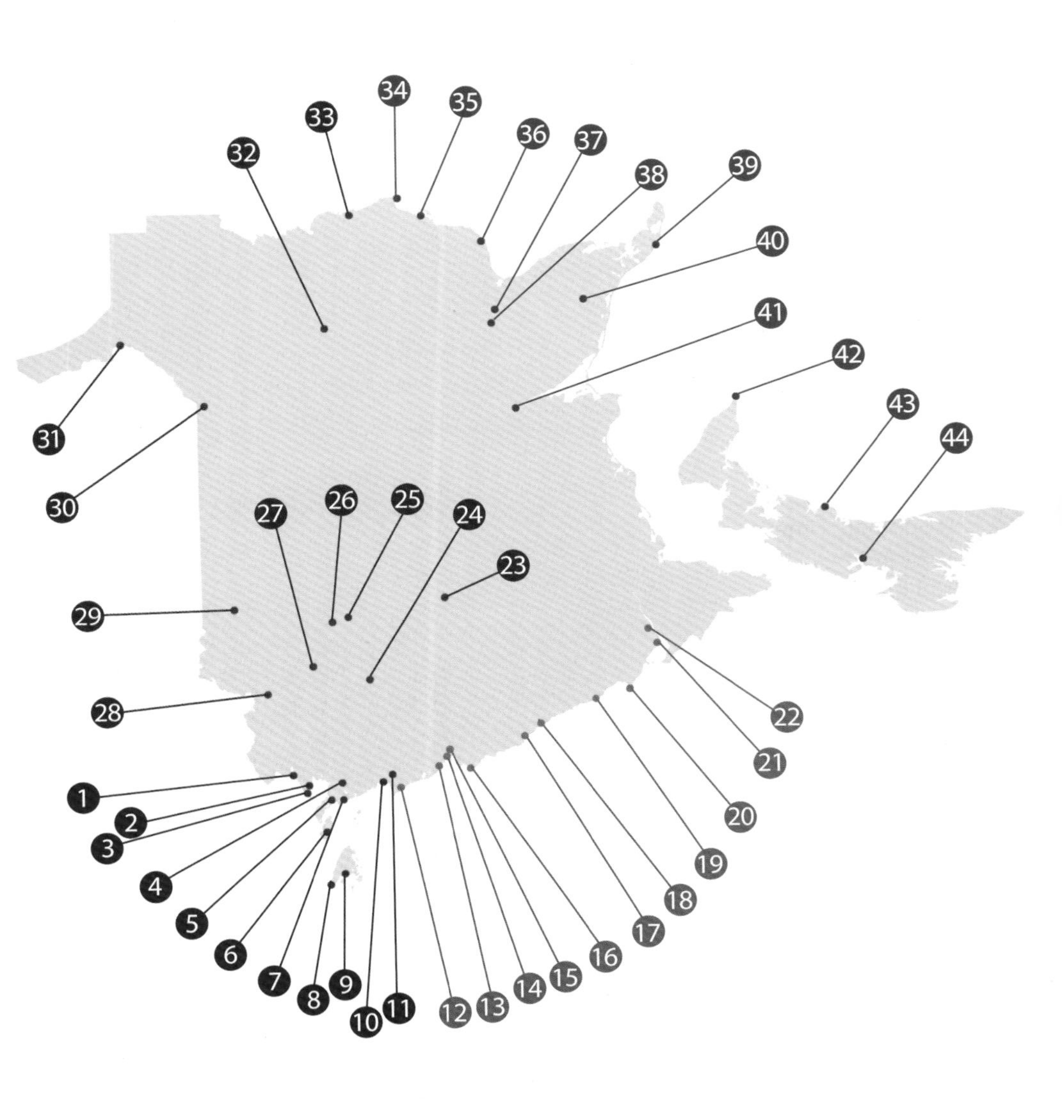
1
2
3
4
5
6
7
8
9
10
11
12
13
14
15
16
17
18
19
20
21
22
23
24
25
26
27
28
29
30
31
32
33
34
35
36
37
38
39
40
41
42
43
44

Table of Contents

Authors' Preface

In New Brunswick and Prince Edward Island, nature paints with bold strokes: high mountains, long rivers, rushing tides, endless beaches. Underlying the landscape—shaping the land and its history—is solid bedrock, the subject of this book. The bedrock itself speaks boldly, too. If you spend a bit of time exploring the rocks here, along with great scenery you'll find dramatic stories of the geologic past.

This book focuses on 44 sites of geological and geoheritage interest in New Brunswick and Prince Edward Island. The site descriptions contain expert information about how the rocks formed, yet the book is not intended for sober-faced study. Its purpose, put simply, is to provide pleasant outings for anyone with a bit of curiosity about the Earth and a penchant for outdoor fun. If you've heard that field geology requires an iron constitution, never fear! We've chosen sites with families and the casual traveller in mind.

Of the 44 sites described here, five focus primarily on the region's rich geoheritage, using stone buildings and other historical resources to showcase the interplay between Earth history and human history. The geoheritage theme is featured in several additional site descriptions as a complement to their geological stories.

Those familiar with the geology field guides for Newfoundland and for Nova Scotia published by Boulder Books will find the same easy-to-use design in this guide. At its core is a series of structured, 4-page site descriptions (6 pages for sites with multiple points of interest) that provide all the information you need to have a good time looking at rocks.

One difference is that the site descriptions are organized around road travel through the region rather than virtual tours through time. The reasons for this are geological. The bedrock map—in New Brunswick especially—is in many areas a crazy patchwork of surprising juxtapositions, with young rocks next to ancient ones, and rocks that were originally thousands of kilometres apart now side by side.

To make sense of the sequence of events, the introduction provides a detailed chronological review of the region's geological history cross-referenced with the site descriptions. The chronological order of site outcrops is also summarized in a simple Time Tour chart.

Natural Resources Canada and the New Brunswick Department of Energy and Resource Development both provide extensive online sources of geological information for this region. Recognizing that some readers may want to explore the rocks in more scientific detail, we have provided features on the third page of each site description that allow quick retrieval of further information.

Complementing our descriptions of 44 sites of interest is a section on highway geology for New Brunswick Routes 1, 2, and 7. Designed for use by vehicle passengers or for virtual touring online, these pages describe the geologic significance and history of rocks exposed along each section of highway and identify selected points of interest.

Welcome to a new way of exploring New Brunswick and Prince Edward Island. From the mountains to the sea, whether you choose to travel by road and trail or in your imagination, the bedrock and its legacy await you. If you use this field guide to explore the region, we hope you'll find as we do that the rocks here tell a fascinating story of Earth history and of society's interactions with Earth resources.

Acknowledgements

The development of this book benefited from the contributions of many individuals. We thank Greg McHone, Mike Parkhill, and Reg Wilson for introducing us to several of the sites and to details of local geology. Les Fyffe and Reg Wilson generously served as expert readers for a draft of the book and answered numerous questions for us on the finer points of regional geology. Sue Johnson, Randy Miller, Adrian Park, Matt Stimson, and Chris White provided professional consultation and advice.

Dave Hild and Alan Macdonald provided unflagging encouragement and practical assistance for many aspects of the project and deserve our special thanks, as does Florence Grant for carefully reviewing our page proofs.

Several individuals and organizations generously granted permission to use images in the book. We thank the *Canadian Journal of Earth Sciences*, Sue Johnson, Marci Miller, Mike Parkhill, Kevin Snair, Richard Veenhuis, the Village of Minto, and Reg Wilson. See the Image Credits in the back of the book for details.

Local residents at several sites helped us track down features and information of interest. We thank the 2018 summer program coordinators at Ganong Nature Park, Todds Point; the pro shop staff at the Hillsborough Golf Club; Minto residents Reggie Barton, Bob Coakley, Wendy Flowers, and Rosalyn Gray; the staff at the McAdam campground; and the staff at the Grand Falls Malabeam Interpretive Centre.

Knowledge about the geology of New Brunswick and Prince Edward Island is continually evolving. Members of the geological surveys of Canada and of New Brunswick, faculty and students of the region's universities, and others who have mapped and studied these provinces' rock formations have all contributed to the information presented here. In particular, we thank Sandra's students over the past 35 years who have shared her enthusiasm and helped her to better know and understand the rocks of southern New Brunswick.

For pulling the whole project together, thanks are due to the patient and talented staff of Boulder Books. who guided the book into print with such attention to detail. For these efforts we are indebted to copy editor Iona Bulgin, designer Tanya Montini, and editor Stephanie Porter.

Financial Support

Grants from the Canadian Geological Foundation and the Atlantic Geoscience Society supported the field collaborations that were so essential to the success of our project. We are indebted to both these organizations for their contributions.

How to Use This Guide

What's in the Guide

This book is organized as a series of four regional tours and aims to include all the "provisions" you'll need along the way. Here's a brief summary.

Sample Pages

Pages 4 and 5 provide a graphical key to the format of the field guide's 44 site descriptions.

Geology Basics

These pages are for readers who want some background information about geology. How do geologists measure and describe geologic time? What are the main kinds of rocks and how do they form? How do tectonic plates interact? Rock types and plate tectonic settings are cross-referenced to specific sites to help provide context.

Regional Background—Five Stories

This section provides a virtual tour through time. Four narratives provide a chronological account of the forces that shaped New Brunswick's and Prince Edward Island's geological history. A fifth reviews elements of the region's geoheritage. All are cross-referenced to specific sites illustrating key elements of the region's geological past and its legacy.

Resources

For the extra curious, following Regional Background is a list of geology-themed museums and interpretive centres as well as print and online resources you can explore to learn more.

Trip Planner and Time Tour

Next is a Trip Planner you can use to select sites and plan your itinerary based on what interests you, what parts of the region you'll be visiting, how far you want to hike, and other preferences. It is followed by a Time Tour, which lists the same sites in order of geological age.

Site Descriptions

The core of the book describes 44 sites of geological interest, organized by geographical region. For each site, photographs, maps, diagrams, and text provide information about why the site is interesting, how to get there, what to look for, and what the outcrop means for the geological history and/or geoheritage of the region.

Highway Geology

This section explains the geological significance of rock outcrops along Routes 1, 2, and 7.

Glossary

Words that may be unfamiliar are explained at the back of the book.

Index

An alphabetical index of place names includes all locations of geological or geoheritage interest mentioned in the text. It contains about 100 entries in all.

Legends

Book Sections

Throughout the book, specific colours are associated with each of the four tours through New Brunswick and Prince Edward Island.

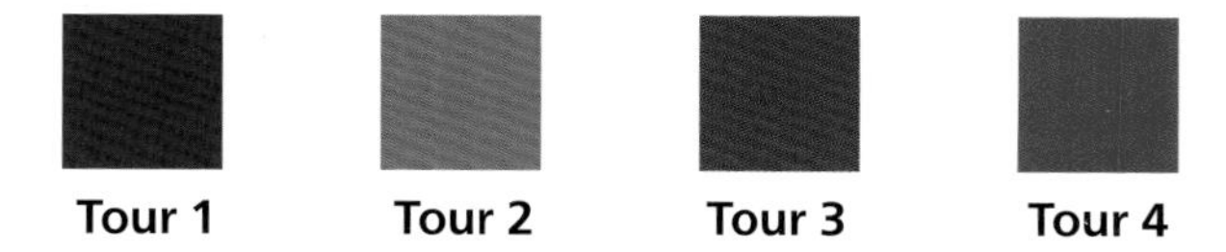

NOTE: All road maps and most geologic maps are oriented with North at the top. In maps with other orientations, the direction of North is indicated.

Map Symbols

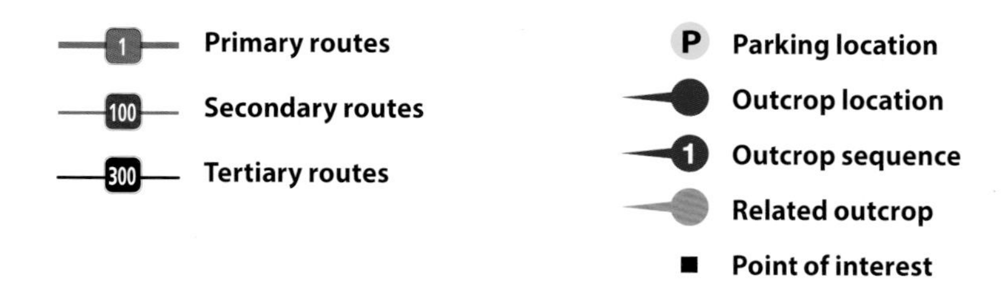

Icons

NOTE: Icons relevant to the site are printed in a dark, solid colour. Icons not applicable are printed in a lighter tone.

Walking distance: This is the approximate total walking distance for the site, that is, the distance from the parking location to the outcrop and back.

Water level: This outcrop is best viewed in low-water conditions (low tide or low lake and river levels). Check tide tables or other sources when planning your visit. NOTE: This icon refers to normal tides and normal seasonal water-level fluctuations only. Storms or other unusual conditions affecting water levels may pose safety and access issues even for sites not marked with this icon.

Park: This site is located within a national, provincial, or community park, or other protected area, and may not be open year-round. It is your responsibility to be aware of park rules and policies affecting your visit.

Cost: There is a fee to visit this site and/or the park in which it is located.

Lighthouse: There is a lighthouse at this site, adding to its scenic value. (But be aware that some lighthouses include foghorns that may sound loudly when visibility is poor.)

Also Note

- For each site, a map number and name are listed under the heading 1:50,000 Map. These identify the National Topographic System of Canada (NTS) map on which the site appears.
- For each site, the formal geological name of the rock unit is listed, along with the reference number of the provincial Bedrock Map of the area. This information can be used to look up further geological details online (see Resources, p. 29).
- All latitude-longitude readings in the field guide are given in the same format, for example, N45.17057 W67.16007. The letter prefix refers to the hemisphere (northern latitude and western longitude) and should be included when entering the coordinates into a GPS device or mapping utility.
- In a few of the photographs, a black-and-white metric scale is visible. The scale is 10 centimetres long; its smallest subdivisions are 1 centimetre square.

Site location in easy-to-reference heading, colour-coded by tour.

Local scenery to help you recognize the site when you arrive.

Geological story represented by the rocks at the site.

Non-technical **description** of the significance and features at the site.

9. THE ANCHORAGE PARK

Long Pond Beach is the site of several interesting outcrops in Anchorage Provincial Park on Grand Manan Island.

A

Faults, Folds, Flows

Ediacaran Rocks of Eastern Grand Manan Island

Geologists often describe Grand Manan Island as having a dual identity. The larger, western part is blanketed in flood basalt (site 8, Southwest Head). On the eastern side is a complicated array of much older rocks, the origins of which are still being debated.

In eastern Grand Manan Island and nearby islands, faults have brought numerous terrane fragments side by side. Some evidence suggests that most were once part of the mainland New River terrane. But even basic age information for some of the rocks has eluded researchers, and mysteries remain (see Also Nearby). If you read other books and papers about them, beware the phrase "assumed to be."

The Anchorage Provincial Park provides easy access to three significant features contributing to the island's complexity: (1) It provides the best available view of the major fault separating west from east. (2) The park's long shoreline exposes elaborately folded Ediacaran sedimentary rocks of mysterious provenance. (3) On its popular sandy beach, outcrops of volcanic rock display a variety of features formed by ash flows and other eruptions, also during the Ediacaran period.

1

On the Outcrop (1)

Groundwater seeping along the fault at Red Point has given rise to a line of small plants along the fault trace (see arrows) dividing basalt (left) from much older rocks (right).

Outcrop Location: N44.64773 W66.82065

At this site, note that the property above the fault is privately owned—please stay on the shore until you can re-enter the park.

No wonder the Red Point fault looks dishevelled. Geologists estimate that the basalt layers on the left slipped down about 2 kilometres along this slanting surface, relative to the ancient, reddish, metamorphosed volcanic rocks on the right. This type of fault, along which the rocks above slide down, is known as a normal fault.

Immediately beside the fault the rocks on both sides are crumbly, broken up as they scraped against one another. About 30 metres west of the fault, its effects can still be seen. Once-vertical columns of basalt were deformed during fault movement and now look twisted or bent (detail). Farther west of the fault the columns are more nearly vertical.

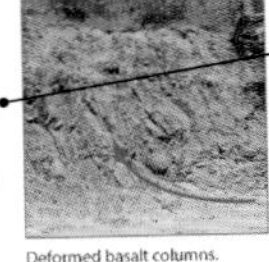

Deformed basalt columns.

Rock Unit	Bedrock Map
Dark Harbour Basalt; Long Pond Bay Formation, Castalia Group	MP 2011-14

3

Image of the outcrop to help you find and interpret the rocks.

GPS waypoint for the outcrop location—in decimal degrees so you can easily enter it into your GPS device or web browser.

Description of the outcrop emphasizing what to look for.

Official **rock unit** name and **bedrock map** publication number, so you can find more technical details if you are curious (see Resources, p. 29).

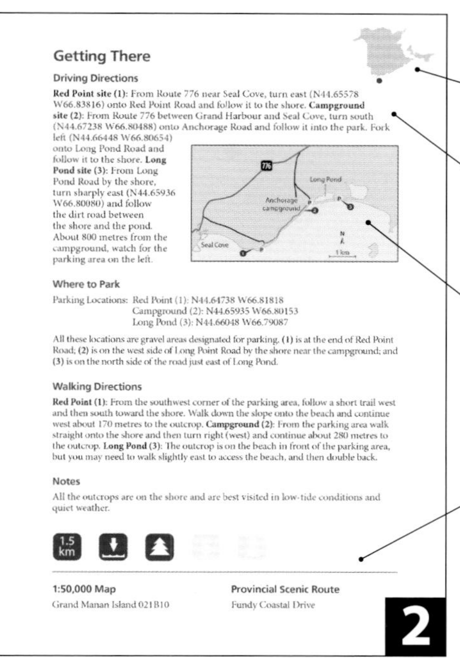

Getting There

Driving Directions

Red Point site (1): From Route 776 near Seal Cove, turn east (N44.65578 W66.83816) onto Red Point Road and follow it to the shore. **Campground site (2):** From Route 776 between Grand Harbour and Seal Cove, turn south (N44.67238 W66.80488) onto Anchorage Road and follow it into the park. Fork left (N44.66448 W66.80654) onto Long Pond Road and follow it to the shore. **Long Pond site (3):** From Long Pond Road by the shore, turn sharply east (N44.65936 W66.80080) and follow the dirt road between the shore and the pond. About 800 metres from the campground, watch for the parking area on the left.

Where to Park

Parking Locations: Red Point (1): N44.64738 W66.81818
Campground (2): N44.65935 W66.80153
Long Pond (3): N44.66048 W66.79087

All these locations are gravel areas designated for parking. **(1)** is at the end of Red Point Road; **(2)** is on the west side of Long Point Road by the shore near the campground; and **(3)** is on the north side of the road just east of Long Pond.

Walking Directions

Red Point (1): From the southwest corner of the parking area, follow a short trail west and then south toward the shore. Walk down the slope onto the beach and continue west about 170 metres to the outcrop. **Campground (2):** From the parking area walk straight onto the shore and then turn right (west) and continue about 280 metres to the outcrop. **Long Pond (3):** The outcrop is on the beach in front of the parking area, but you may need to walk slightly east to access the beach, and then double back.

Notes

All the outcrops are on the shore and are best visited in low-tide conditions and quiet weather.

1.5 km

1:50,000 Map
Grand Manan Island 021B10

Provincial Scenic Route
Fundy Coastal Drive

2

Thumbnail map showing site location.

Detailed written **directions** for driving, parking (including GPS waypoint), and walking.

Road map showing the route to the outcrop from the nearest highway (for Legend, see p. 2).

At-a-glance **information** to help you plan your trip: key site characteristics plus map grid and tourism route details.

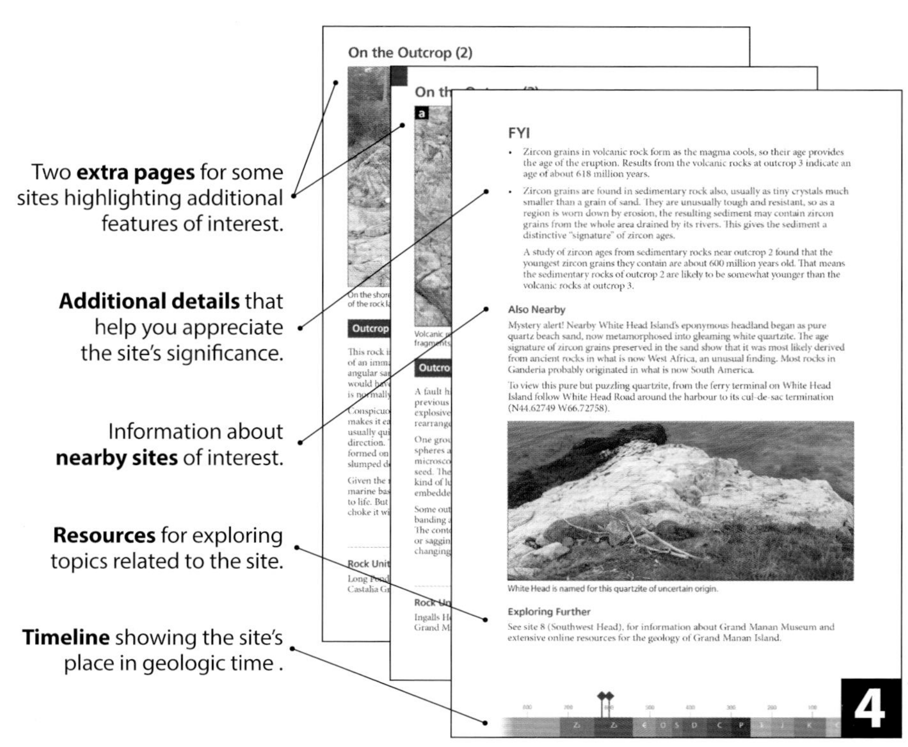

On the Outcrop (2)

FYI

- Zircon grains in volcanic rock form as the magma cools, so their age provides the age of the eruption. Results from the volcanic rocks at outcrop 3 indicate an age of about 618 million years.
- Zircon grains are found in sedimentary rock also, usually as tiny crystals much smaller than a grain of sand. They are unusually tough and resistant, so as a region is worn down by erosion, the resulting sediment may contain zircon grains from the whole area drained by its rivers. This gives the sediment a distinctive "signature" of zircon ages.

  A study of zircon ages from sedimentary rocks near outcrop 2 found that the youngest zircon grains they contain are about 600 million years old. That means the sedimentary rocks of outcrop 2 are likely to be somewhat younger than the volcanic rocks at outcrop 3.

Also Nearby

Mystery alert! Nearby White Head Island's eponymous headland began as pure quartz beach sand, now metamorphosed into gleaming white quartzite. The age signature of zircon grains preserved in the sand show that it was most likely derived from ancient rocks in what is now West Africa, an unusual finding. Most rocks in Ganderia probably originated in what is now South America.

To view this pure but puzzling quartzite, from the ferry terminal on White Head Island follow White Head Road around the harbour to its cul-de-sac termination (N44.62749 W66.72758).

White Head is named for this quartzite of uncertain origin.

Exploring Further

See site 8 (Southwest Head), for information about Grand Manan Museum and extensive online resources for the geology of Grand Manan Island.

4

Two **extra pages** for some sites highlighting additional features of interest.

Additional details that help you appreciate the site's significance.

Information about **nearby sites** of interest.

Resources for exploring topics related to the site.

Timeline showing the site's place in geologic time .

Exploring the Sites

Safety

You can't experience geology directly without going outdoors. All the sites in this book have been visited safely by many people. They are places where residents, tourists, geologists, students, and others go to enjoy the beauty of the region and examine interesting rocks. However, weather, tides, and other conditions can render any site hazardous temporarily. Only you can decide whether it is safe to visit a specific site on a specific day and whether your state of preparedness is appropriate for the conditions.

Navigating

Each site description provides information to help you find the outcrop easily. Road maps and written directions indicate one possible route from the nearest highway to the parking location; further instructions describe how to reach the outcrop. Additional details help you find the site on topographic or tourism maps.

All field readings of GPS latitude and longitude reference World Geodetic System datum WGS84. If you enter them into your own GPS device, web browser, or other application, keep in mind that results may vary depending on your GPS device settings, map projections, application preferences, and other factors. The readings are not intended, and should never be used, as a substitute for attentive real-world navigation.

The parking locations given are example locations where parking was possible at the time of publication. Please exercise common sense and courtesy at all times when parking your vehicle, based on the conditions you encounter. The listed outcrop location marks the position of the rock described in the text or a vantage point from which significant features can be viewed.

Written directions to the sites include words and phrases such as "northward," "to the east," or "southwest." They are used only in a general sense to clarify correct choices at forks, intersections, and other turning points. They are not intended as precise orienteering instructions. Driving and hiking distances are approximate and are provided only as an aid in planning your trip.

Preserving Outcrops

The geology and geoheritage of New Brunswick and Prince Edward Island are unique. If you use this field guide to visit sites, please apply the principles of Leave No Trace Canada (www.leavenotrace.ca) to preserve the provinces' geological and geoheritage treasures for others to enjoy. Don't hammer outcrops or remove material—photographs are the best way to capture your experiences.

Protecting Fossils

In both New Brunswick and Prince Edward Island, the fossil record is protected by law. Under New Brunswick's *Heritage Conservation Act*, the province owns all fossils (including fossil tracks) that have been or may be discovered. It is against the law to damage, destroy, or remove any fossils from sites where they are found without a permit issued by the province. Similar policies apply on Prince Edward Island, where the fossil record is protected by the province's *Heritage Places Protection Act*.

Geology Basics

Geologic Time

Superposition and Cross-Cutting

The features of an outcrop can reveal the relative ages of geologic events, that is, the order in which they occurred. The principles of superposition and cross-cutting relations allow you to reconstruct a sequence of events even though they happened long ago.

Superposition. Sedimentary rocks are deposited in horizontal layers, with younger layers on top of older layers.

Cross-cutting. Any feature (intrusion, fault, erosion surface) that cuts across or truncates other features is younger than all the features it cuts across.

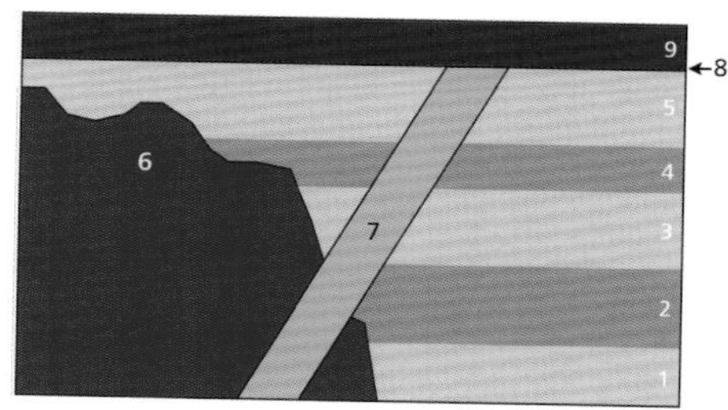

A simplified rock outcrop showing the order in which features formed: 1–5, sedimentary layers; 6, igneous intrusion; 7, cross-cutting dyke; 8, erosion surface; 9, a younger sedimentary layer.

Fossil Species and Radiometric Data

Geologists use two types of evidence—fossil species and radiometric data—to tell when rocks formed and geologic events happened.

Fossil species. As life evolved in the geologic past, new species appeared, existed for a time, and became extinct. Based on close study of the appearance, distribution, and extinction of fossil species preserved in rock, the international geological community has agreed on standards that define which fossils belong to which geologic period. This information allows a rock layer to be assigned an age based on the fossils it contains.

Radiometric data. Some rocks and minerals contain small amounts of radioactive atoms. Radioactive decay converts an unstable atom (called the parent isotope) into a stable atom (called the daughter isotope). For example, radioactive atoms of uranium decay to form stable atoms of lead. By measuring the amount of parent and daughter isotopes very precisely, geologists can calculate the age of the rock based on the known rate of radioactive decay.

Proterozoic Eon

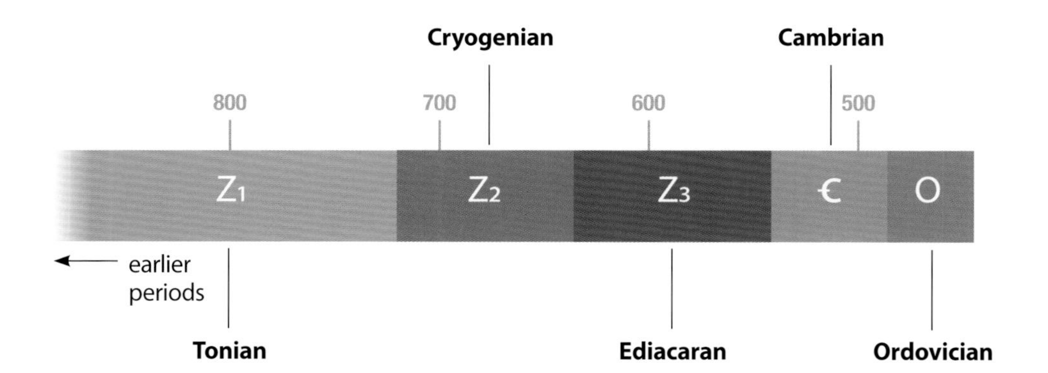

The Book's Timeline*

On the fourth page of each site description is a ribbon of colour like the one shown above. The numbers on the timeline mark off intervals of 100 million years before the present day; the colours mark the boundaries of geologic periods and eras; and the letters are abbreviations for the names of the periods (except Cenozoic, which is an era, not a period).

Eras are shown on the timeline by colours: shades of brown for the Neoproterozoic (or late Proterozoic) era; shades of blue for the Paleozoic era; shades of green for the Mesozoic era; and yellow for the Cenozoic era.

Each site description includes a marker on the timeline showing the age(s) of the rock(s) or event(s) of interest at that site. Events described in this book span a period of nearly half a billion years, from 700 to 200 million years ago. In the example above, the marker is placed at 360 million years ago.

Exploring Further

Canadian Federation of Earth Sciences. *Four Billion Years and Counting*, Ch. 3, "It's about Time" (pp. 38–44).

---

* Keen-eyed readers of Boulder's *Geology of Newfoundland* or *Geology of Nova Scotia* will notice new time boundaries for some geologic periods, especially the Cryogenian. The International Commission on Stratigraphy made these changes in 2015.

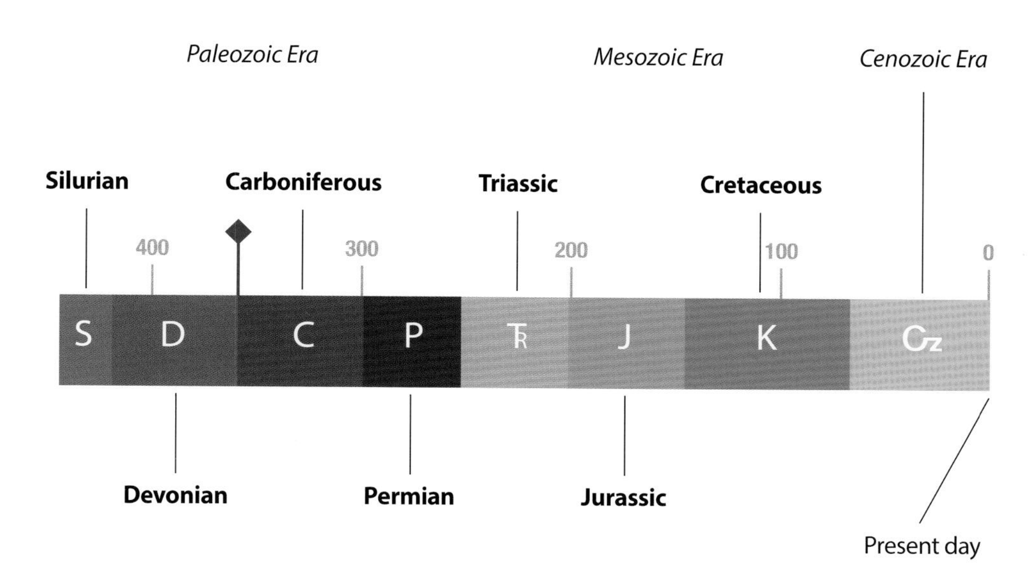

Four Eons of Geologic Time

The book's timeline shows only part of the history of the Earth—our planet formed about 4,550 million years ago. Geologists divide geologic time into eons, which are subdivided into eras and periods. Here's a quick summary of what happened during each eon of geologic time:

Hadean eon (4,550 to 4,000 million years ago). Very few rocks or minerals survive from the Hadean eon. The Earth's crust was hot, mobile, and bombarded by meteorites.

Archean eon (4,000 to 2,500 million years ago). Tectonic plates and early continents formed; primitive single-celled micro-organisms appeared in the sea.

Proterozoic eon (2,500 to 541 million years ago). Areas of stable continental crust grew and mountain belts formed during a series of continental collisions. Oxygen levels increased in the air and sea, allowing early complex life forms to evolve.

Phanerozoic eon (541 million years ago to the present). Continents re-assembled to form the supercontinent Pangaea. Pangaea broke up as present-day continents moved to their current positions. Animals and plants evolved to inhabit both sea and land.

Rock Types

Identifying Rocks

New Brunswick and Prince Edward Island provide access to a variety of rock types, each one a doorway to a fascinating story from the Earth's past. Correctly identifying the type of rock that you are looking at is the first step in exploring its history.

Plenty of general guides to rock identification are available in print and online. This brief review uses examples from the sites described in this book.

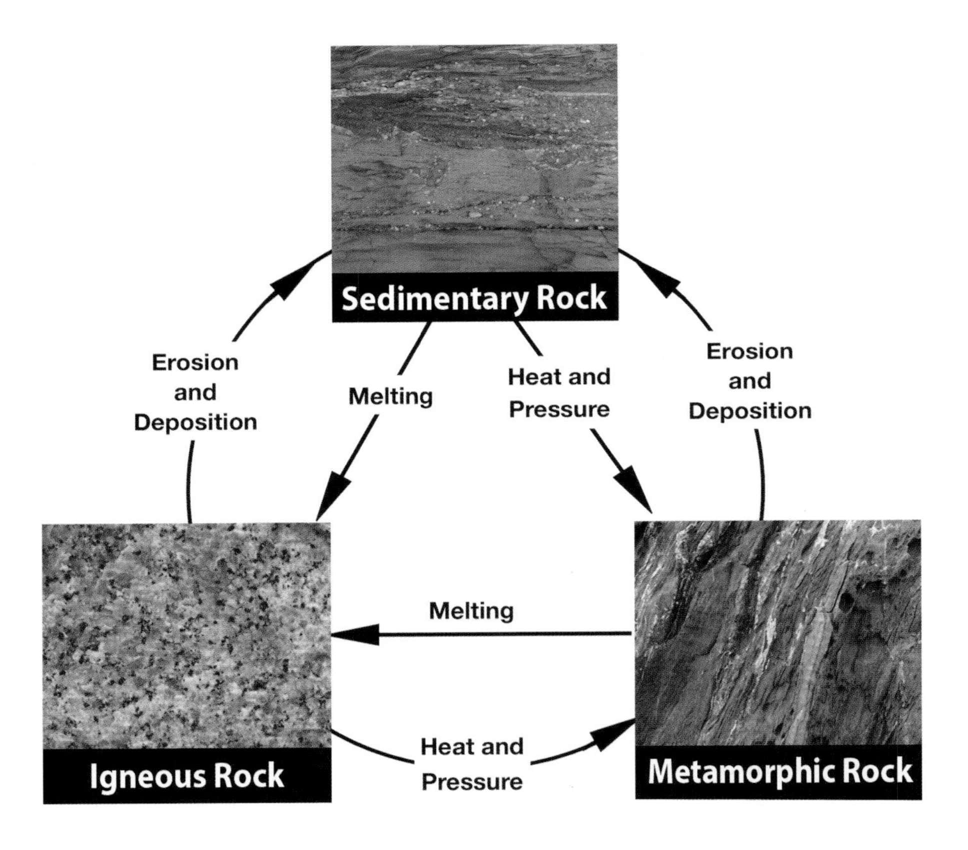

The rock cycle shows how igneous, sedimentary, and metamorphic rocks can be transformed and recycled into other rock types when exposed to new conditions. Changing conditions can be caused by uplift, burial, the movement of fluids in the crust, and other events as the Earth's tectonic plates interact.

Igneous Rocks

Igneous rocks form when molten rock (magma) rises through the crust, cools, and crystallizes. Igneous rocks are classified based on the minerals they contain, then subdivided into two varieties: intrusive (crystallized slowly underground and thus coarse-grained) and extrusive (erupted onto the surface, crystallizing quickly and thus fine-grained).

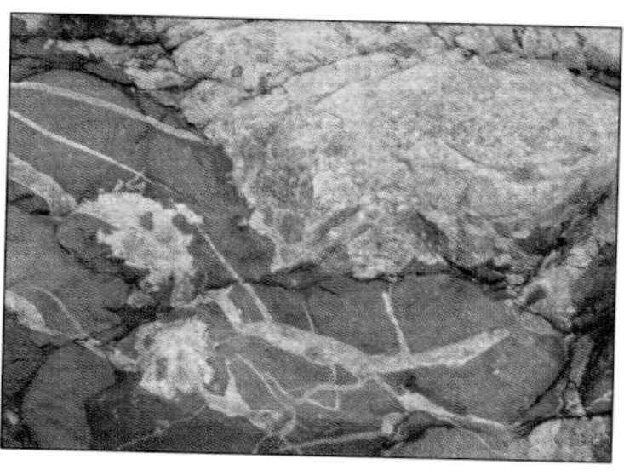

Granite and basalt, Todds Point (site 1).

Felsic igneous rocks include granite (intrusive—sites 1, Todds Point; 4, St. George; 16, Cape Spencer; 28, McAdam; 37, Pabineau Falls) and rhyolite (extrusive—sites 9, The Anchorage Park; 12, Dipper Harbour; 18, Fundy Trail Parkway; 19, Point Wolfe; 27, Harvey; 32, Williams Falls). They contain mostly light-coloured minerals (quartz and feldspar), typically with small amounts of dark minerals such as biotite or amphibole. True granite contains potassium feldspar, plagioclase, and quartz. The term felsic is used to refer to a wide variety of igneous rocks dominated by any combination of these light-coloured minerals.

The most common intermediate rocks (sites 10, Barnaby Head; 15, Rockwood Park) are granodiorite and diorite (intrusive) or their fine-grained equivalents, dacite and andesite (extrusive). They contain a mixture of light and dark minerals.

Mafic igneous rocks (sites 1, Todds Point; 3, Ministers Island; 8, Southwest Head; 13, Taylors Island; 25, Currie Mountain; 36, Atlas Park) contain approximately equal amounts of plagioclase and dark minerals, usually pyroxene. The plagioclase is calcium-rich and typically grey. Gabbro (intrusive) and basalt (extrusive) both have a similar mixture of minerals but differ in the size of the mineral grains—easily visible in gabbro and microscopic in basalt.

Some igneous rocks do not fit this simple classification. Trachyte, for example, is made almost exclusively of potassium feldspar (sites 33, Sugarloaf; 34, Inch Arran Point), but contains little or no quartz. Subvolcanic rocks (sites 25, Currie Mountain; 32, Williams Falls, outcrop 1; 33, Sugarloaf) are intrusive but formed so close to the surface that they share many characteristics of extrusive rocks.

Sedimentary Rocks

Sedimentary rocks typically form in layers. In many kinds, the layers originate as loose sediment, most commonly deposited underwater. They harden into rock as they become deeply buried by more sediment.

Most sedimentary rocks in New Brunswick and Prince Edward Island are made of broken-up, weathered fragments eroded from older rocks. The particle size of the sediment determines the rock name. Tiny clay particles form mudstone or shale; sand forms sandstone (or, if its grains are very small, siltstone); and gravel and larger fragments form conglomerate.

Sedimentary rocks formed along continental margins and in deep ocean basins typically consist of alternating layers of sandstone or siltstone and shale (sites 14, King Square West; 29, Hays Falls; 30, Grand Falls; 31, Edmundston; 38, Middle Landing). Along steep underwater slopes turbidites may form (sites 6, Herring Cove; 26,

Mactaquac Dam). They are deposited during specific events in which a sediment-laden current rushes down a steep underwater slope, sometimes triggered by earthquakes.

On land, rivers move sediment from highlands into lower-lying areas. Steep slopes, for example beside active faults, create energetic flows of water (sites 2, St. Andrews; 7, Pea Point; 17, St. Martins; 21, Hopewell Rocks) that can transport gravel-sized sediment or even boulders. In areas of gentler slope, rivers deposit sand, silt, and mud in the riverbed and adjacent flood plains (for example, sites 20, Cape Enrage; 24, Fredericton Junction; 41, French Fort Cove). If abundant plant material accumulates, burial by younger sediment may transform the plant remains into coal (site 23, Minto).

In hot, dry continental regions with infrequent rain, rivers may run dry but also may flood after violent or prolonged storms. Sedimentary rocks formed in this setting tend to have a characteristic red or orange colour and are known as continental red beds (sites 11, Lepreau Falls; 17, St. Martins; 39, Cap-Bateau; 42, North Cape; 43, Cavendish Beach). The individual particles of sediment are not red. Instead, as they are deposited and buried under new layers of sediment, each particle becomes coated with colourful ferric iron oxide carried by circulating groundwater.

Sandstone, Ministers Island (site 3).

Iron oxide takes two forms, depending on environmental conditions. Reddish ferric oxide (Fe_2O_3) forms in oxygen-rich environments with neutral pH (neither acid nor base). When less oxygen is available or pH is altered, dark grey ferrous oxide (FeO) forms instead. The decay of organic matter uses a lot of oxygen and can result in grey layers or patches in an otherwise red or orange sedimentary rock (sites 17, St. Martins; 42, North Cape).

Some sedimentary rocks do not contain weathered pieces of older rock, but instead form chemically when minerals precipitate from sea or lake water, in some cases with the help of organisms. Carbonate rocks like limestone form in this way. Gypsum (site 22, Hillsborough) forms when sea water evaporates in extreme conditions—for example in a shallow tropical sea.

Metamorphic Rocks

Metamorphic rocks form when existing rocks of any type recrystallize due to changes in temperature and/or pressure—for example, when a region of the Earth's crust is buried under a colliding tectonic plate. Metamorphic rocks are classified based on the intensity of the metamorphism, which is inferred from the presence of specific metamorphic minerals.

Marble, Rockwood Park (site 15).

Carboniferous and younger rocks in New Brunswick and Prince Edward Island have never been deeply buried, so they have been little affected by metamorphism. However, the older rocks of New Brunswick include metamorphic rocks such as marble, quartzite, and slate (sites 15, Rockwood Park; 29, Hays Falls; 31, Edmundston; 38, Middle Landing), formed from limestone, sandstone, and shale respectively.

Folds and Faults

During tectonic plate interactions, rocks can be folded as crustal blocks move toward or slide past one another. When this happens, flaky metamorphic minerals like mica may become aligned, resulting in foliated rock types that split apart easily (sites 31, Edmundston; 32, Williams Falls; 38, Middle Landing). Where folds have formed on a large scale, their whole shape may not be visible in a single outcrop, and the only indication of folding is that the rock layers are tilted, as seen in many rocks of southern and western New Brunswick. Intense deformation during fault movements can create characteristic broken and sheared rock textures such as mylonite (sites 10, Barnaby Head; 16, Cape Spencer).

Fault movements also play an important role in shaping the landscape, in places producing dramatic, linear breaks in topography or steep-sided valleys. New Brunswick and Prince Edward Island are criss-crossed by innumerable faults, many marking abrupt changes in rock type and geological history (sites 9, The Anchorage Park; 15, Rockwood Park) and others affecting the location of valleys (see Highway Geology).

Layers tilted by folding, Grand Falls (site 30).

Exploring Further

Bishop, Arthur C., Alan R. Wooley, and William R. Hamilton. *Guide to Minerals, Rocks & Fossils.* Firefly Books Ltd., 2005.

Canadian Federation of Earth Sciences. *Four Billion Years and Counting*, Ch. 1, "On the Rocks" (pp. 4–18).

Plate Tectonics

Without an understanding of plate tectonics, it's unlikely geologists would ever have made sense of this region's complex geological past. Here you'll find some of the ideas and terms geologists use when thinking and writing about tectonic plates.

Tectonic Plates

Since at least 2,500 million years ago, the Earth's outermost layer, the lithosphere, has been cool and rigid enough to form well-defined tectonic plates that include oceanic crust, continental crust, or both. The plates are carried around, pulled apart, and pushed together in a variety of ways by convection in the mantle below.

The two types of crust are very different. Oceanic crust is relatively thin and dense. It forms where plates are being pulled apart; and it is destroyed—by sinking back into the mantle—where plates are being pushed together. In consequence, oceanic crust never gets very old. The maximum age of the crust in today's oceans is only about 200 million years.

Continental crust is thicker and less dense than oceanic crust. It forms through complex processes when ocean crust sinks back into the mantle (see Convergent Boundaries, below). The low density of continental crust causes it to float on the mantle like ice in a cool drink. Due to its buoyancy, continental crust may be changed but will never sink into the mantle, so the continents preserve rocks of many different ages—from ancient to modern.

Plate Interactions

The boundary between two plates can be called divergent, convergent, or transform, based on how the plates interact. Along present-day plate boundaries, distinctive patterns of earthquake and volcanic activity reveal what's going on. To understand plate motions of the past, geologists look for clues in the rocks. Each type of plate interaction leaves a "signature" of rock features.

Divergent Boundaries

At divergent plate boundaries, hot regions of the mantle flow upward and then away from the plate boundary. Shallow earthquakes occur as the Earth's crust cracks, molten rock rises to fill the void, and the plates move apart.

When divergent motion affects an area of continental crust, continental rifting occurs. In New Brunswick and Prince Edward Island, dykes mark the site of crustal breaks (site 3, Ministers Island) associated with the early stages of formation of the Atlantic Ocean. As continental rifting continued, a steep-sided valley formed and was filled with sedimentary layers (site 17, St. Martins) and great volumes of lava (site 8, Southwest Head).

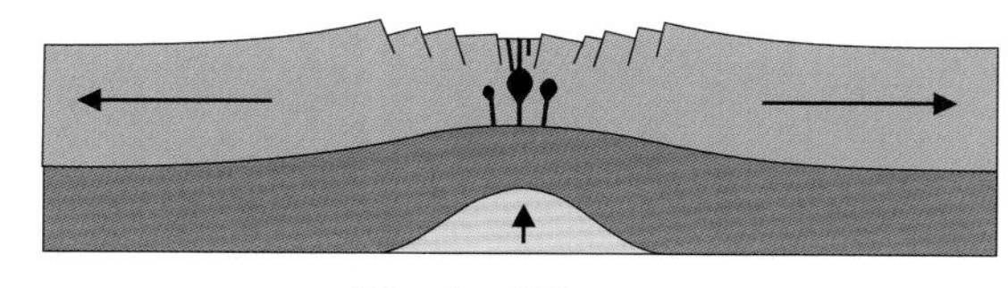

Early Stage

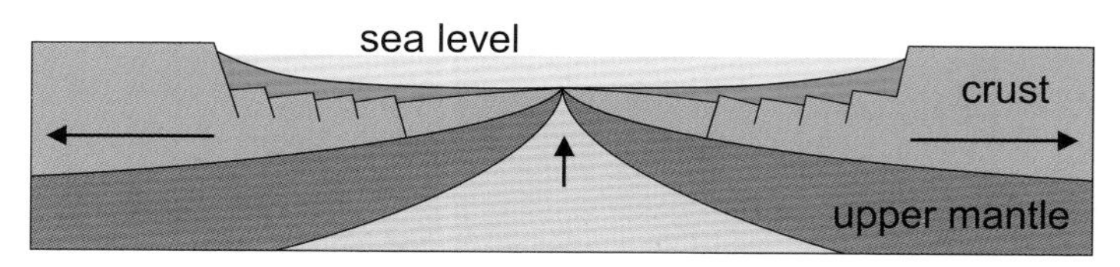

Later Stage

Divergent motion over a long period eventually leads to the formation of oceanic crust in a widening ocean tract between continents. New Brunswick includes small remnants of oceanic crust, caught up in a later collision (site 36, Atlas Park) and preserved.

As a tract of ocean widens, tectonic activity is confined to the mid-ocean ridge, while the continental margins are quiet (passive). New Brunswick preserves rocks that formed on Ganderia's passive margins during the Ediacaran and Cambrian periods (sites 15, Rockwood Park; 29, Hays Falls; 38, Middle Landing). Atlantic Canada's offshore regions are themselves a present-day passive margin.

Convergent Boundaries

Where cool regions of the mantle flow toward one another, then down into the Earth, a convergent boundary forms, causing plate collisions. Convergent boundaries cause deep-seated, severe earthquakes and the Earth's most extreme topography—high mountains as well as large basins.

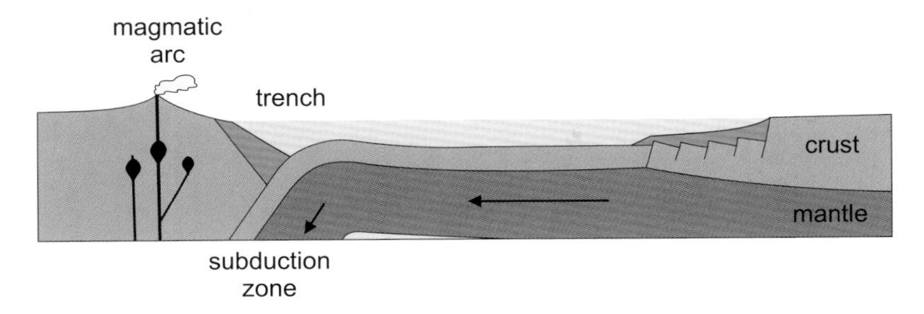

New convergent boundaries nearly always form within older, cooler regions of ocean crust. The plate tears as one edge sinks, forming a deep ocean trench. A process known as subduction carries the sinking plate down into the mantle. Ocean crust on the seafloor contains a certain amount of water, but during subduction, heat and pressure drive the water out and up into the surrounding mantle, which then partially melts. The resulting magma moves upward, forming intrusions or volcanic activity. Subduction under oceanic crust typically forms an arc-shaped chain of volcanic islands (an island arc) above the lower plate. But subduction under the edge of a continent forms a chain of volcanoes on land, as well as underlying igneous intrusions (a magmatic arc; sites 1, Todds Point; 4, St. George; 10, Barnaby Head; 12, Dipper Harbour; 15, Rockwood Park; 16, Cape Spencer; 19, Point Wolfe).

Convergent plate boundaries may ultimately destroy an entire ocean basin, bringing continents into collision. When that happens, one block of continental crust may ride partly over the other, thickening the crust in the collision zone. The lower crust may melt, sending granitic magma upward to form igneous intrusions (sites 28, McAdam; 37, Pabineau Falls).

Transform Boundaries

At transform boundaries, tectonic plates slide sideways past one another. Earthquakes occur in shallower regions as the cool, brittle crust breaks along faults. In deeper regions, transform motion distorts the hotter, more pliable crust to form rocks such as mylonite (sites 10, Barnaby Head; 16, Cape Spencer).

If transform motion occurs over a long period of time, crustal blocks can be rearranged over long distances. Much of southern New Brunswick appears to have been assembled in this way.

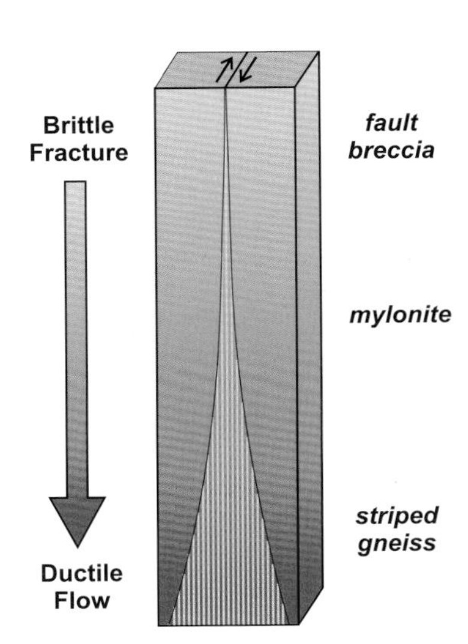

Clues to Past Plate Interactions

Like a couple of retired adventurers, New Brunswick and Prince Edward Island these days are just riding quietly around on the North American tectonic plate. But evidence of plate interactions—in the form of terranes, basins, and major faults—have helped geologists understand the provinces' dramatic past.

Terranes

When mapping a mountain belt, geologists sometimes recognize a discrete area of continental crust with its own distinctive geological features (for example, rock types, metamorphic grade, chemical attributes) and history. They use the word "terrane" to identify and talk about such an area.

Plate interactions can produce small pieces of continental crust, either by rifting off part of a larger continent to form a microcontinent or by generating new continental crust in island arcs. If continents converge over a long period, island arcs, microcontinents, and other landmasses in the intervening ocean basin will eventually collide with continental crust in a process called terrane accretion. In this way, numerous unique bits and pieces of crust have become embedded in the Earth's mountain belts, including the Appalachians of this region (see Regional Background, Story A).

Basins

Basins form an important and fascinating part of the geologic record. By preserving sequences of sedimentary (and in some places volcanic) rock from adjacent regions, basins provide geologists with information about when mountains rose, what the climate, landscape, and life forms were like, how sea level changed, and other clues to help us envision the events and environments of long ago. Fortunately New Brunswick and Prince Edward Island provide quite a showcase of basin types.

Where plates converge, basins can develop in association with volcanic arcs, either forming in front of the arc, as a fore-arc basin (site 30, Grand Falls), or behind it, as a back-arc basin (sites 5, Greens Point; 6, Herring Cove).

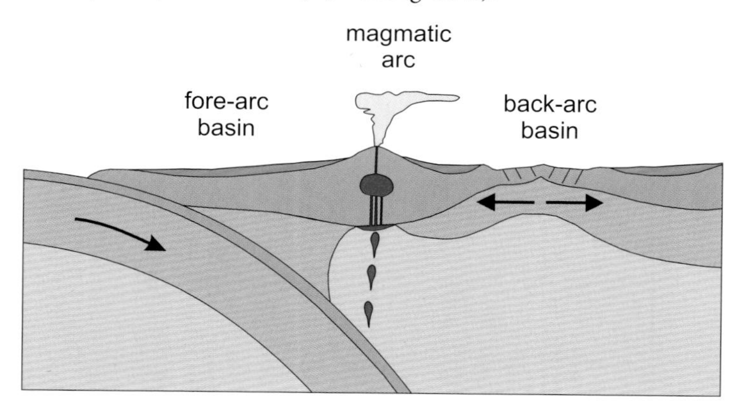

In later stages of convergence, when one slab of continental crust rides up onto another, the mass of the overriding plate warps the underlying crust downward, forming a foreland basin behind the site of collision (site 31, Edmundston).

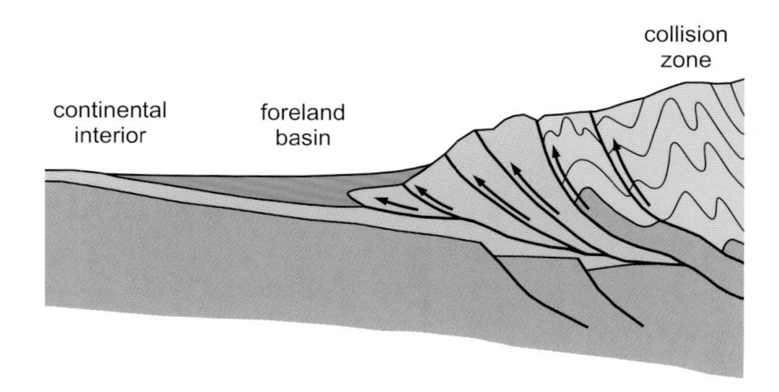

Transform motion can rupture the crust in complicated patterns, forming deep basins between areas of uplift along faults (sites 2, St. Andrews; 7, Pea Point; 21, Hopewell Rocks) or prompting igneous activity marking the site of the rupture (sites 13, Taylors Island; 25, Currie Mountain; 27, Harvey). Many geologists think the large Maritimes basin, which includes much of eastern New Brunswick and all of Prince Edward Island, was formed through this process operating on a large scale.

Major Faults

It's common for large mountain belts to be sliced by numerous long faults. Collisions that continue for millions of years can place tremendous stress on continental crust, even far from the site of collision. Within an evolving mountain belt, pre-existing faults may be reactivated, or new ones formed, to allow a chaotic reshaping of the crust as fault-bounded fragments move—in some cases hundreds of kilometres—in response to the stress. Afterward, the resulting geologic map can appear quite bewildering, a collage of fractured and rearranged pieces. Much of New Brunswick, especially in the south, has been affected in this way.

Exploring Further

Canadian Federation of Earth Sciences. *Four Billion Years and Counting*, Ch. 2, "Dance of the Continents" (pp. 24–37).

Kious, W. Jacquelyne and Robert I. Tilling. *This Dynamic Earth: The Story of Plate Tectonics* (Online Edition). U.S. Geological Survey website, pubs.usgs.gov/ gip/ dynamic. (An excellent summary with numerous illustrations, it can be viewed online or downloaded as a PDF.)

Smithsonian Institution. *Eruptions, Earthquakes, & Emissions* (A time-lapse animation of volcanic eruptions and earthquakes since 1960). Smithsonian Institution website, http://volcano.si.edu/.

Regional Background: Five Stories

Geological Stories

The geological events that shaped this region were complex, and so are the rocks that resulted from them. But thinking about the rocks in terms of just a few basic stories can help bring their significance into focus. Some stories slightly overlap in time, but each has a distinct emphasis.

The bedrock of New Brunswick and Prince Edward Island can be understood in terms of the region's four geological stories. A, Foundations: Ganderia and Avalonia; B, Collisions: Belts and Basins; C, Assembly of Pangaea: The Maritimes Basin; and D, Atlantic: The Fundy Basin. A fifth story, H, Geoheritage, chronicles human interaction with Earth resources.

A. Foundations		B. Collisions		C. Pangaea	D. Atlantic
Ganderia	Avalonia	Basins	Intrusions	Maritimes basin	Rifting

The Big Picture

A billion years ago, the Earth had just one giant landmass, a supercontinent known as Rodinia. Beginning around 750 million years ago, a series of rifts and plate motions broke the supercontinent apart and reorganized the resulting continental fragments. Eventually, by about 540 million years ago, another relatively simple arrangement emerged—just two large continents, Laurentia and Gondwana, separated by a widening sea floor, the Iapetus Ocean.

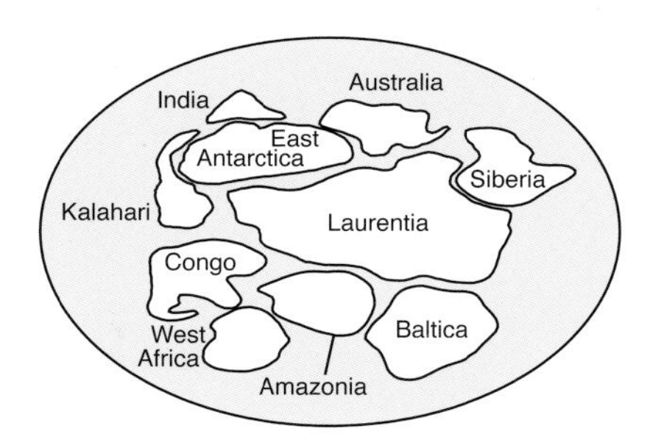

Rodinia.

Soon (in geological time) the picture again became complicated. Along the edge of Gondwana, rifting split off fragments of continental crust—microcontinents—while forming a second ocean region known as the Rheic Ocean. New continental crust in the form of volcanic arcs emerged above subduction zones. Some of the volcanic arcs then rifted, creating additional areas of ocean floor.

Somewhat like the collage of land and sea between Asia and Australia today, the Earth's surface between Laurentia and Gondwana became a complex, ever-changing ocean region littered with landmasses large and small.

In a long sequence lasting hundreds of millions of years, the microcontinents and volcanic arcs collided—sometimes with one another, but ultimately with Laurentia in a process of terrane accretion. Eventually Gondwana itself collided with Laurentia and its accreted terranes, forming a new supercontinent, Pangaea, about 270 million years ago.

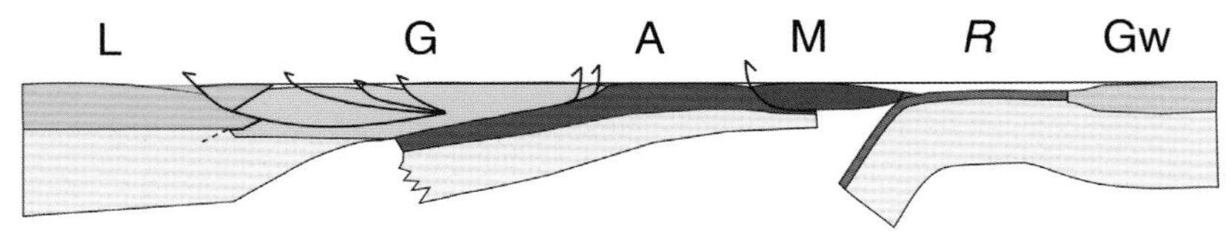

Schematic cross-section showing Laurentia (L), Ganderia (G), Avalonia (A), Meguma (M), the Rheic Ocean (R), and Gondwana (Gw).

With each collision, mountains rose. Huge faults sliced through the region and moved giant slabs of crust up, down, and sideways in vast, bold rearrangements. Rivers relentlessly eroded the high ground and buried low-lying areas in sediment. Just when all seemed quiet at last, about 200 million years ago, the whole story started again like a planetary moment of déjà-vu: Pangaea began to break apart to form the Atlantic Ocean.

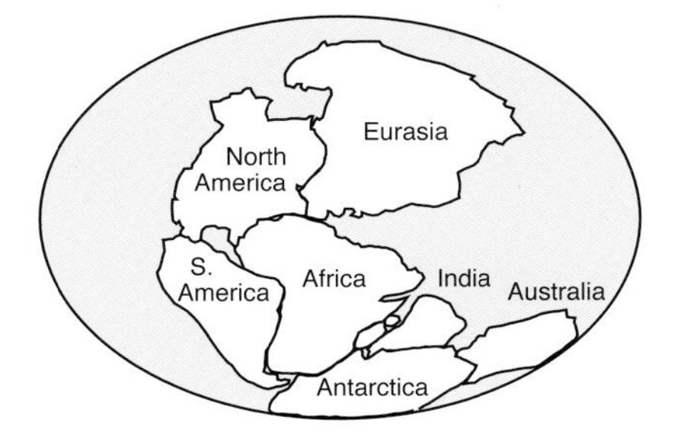

Pangaea.

Foundations: Ganderia and Avalonia

Cryogenian to early Ordovician

No remnants of ancient Laurentia are exposed in New Brunswick or Prince Edward Island. The oldest rocks exposed in the region were part of two microcontinents, Ganderia and Avalonia, that split off from Gondwana. They survive as fault-bounded fragments (terranes) that were rearranged during the Carboniferous period.

Ganderia. Most of New Brunswick originated as part of Ganderia. In fact, New Brunswick preserves the story of Ganderia more fully than any other part of Atlantic Canada. Some New Brunswick rocks formed when Ganderia was still part of Gondwana's margin about 650 million years ago. Along the continental margin in a warm, shallow region of the Iapetus Ocean, limestone and sandstone layers accumulated peacefully (site 15, Rockwood Park).

That quiet setting was disrupted, first about 620 million years ago (site 9, The Anchorage Park), and again about 555 million years ago (sites 10, Barnaby Head; 12, Dipper Harbour; 15, Rockwood Park) when plate motions changed, forming subduction zones beneath Ganderia's margin, causing volcanic eruptions and intrusions.

Rockwood Park (site 15).

The separation of Ganderia from Gondwana took a long time, from about 540 to 490 million years. Large tracts of sandstone and siltstone accumulated as Ganderia's mountainous volcanic regions were eroded into adjacent seaways. Rocks from this time period are preserved in a wide swath between Bathurst and Woodstock (sites 29, Hays Falls; 38, Middle Landing).

By about 490 million years ago, Ganderia was a completely separate microcontinent with the Iapetus Ocean on its "leading edge" (facing Laurentia) and the Rheic Ocean on the "trailing edge" (facing Gondwana).

Tetagouche back-arc basin. Around 480 million years ago, the complex pattern of subduction that was closing the Iapetus Ocean rifted off a slice of Ganderia and opened a new tract of ocean floor that temporarily divided Ganderia: the Tetagouche back-arc basin (site 36, Atlas Park). Volcanic activity on this sea floor led to the mineral deposits of the Bathurst Mining Camp (site 38, Middle Landing, Also Nearby). The basin later closed during the Salinic orogeny (see Story B).

Avalonia. Rocks that originated as part of the microcontinent Avalonia dominate the Caledonia Highlands in southeastern New Brunswick. Its rocks are similar in age to those of Ganderia, but they formed along a different part of the margin of Gondwana.

Avalonia's oldest known rock in New Brunswick is a fragment of ocean floor formed near the margin of Rodinia nearly 700 million years ago (site 18, Fundy Trail Parkway). More abundant and characteristic of Avalonia are volcanic arc rocks and related intrusions with ages of about 620 million years (sites 16, Cape Spencer; 19, Point Wolfe), formed along Avalonia's continental margin while it was still attached to Gondwana.

By around 555 million years ago, tectonic forces were pulling the volcanic arc apart, leading to igneous activity of a different variety (sites 15, Rockwood Park; 18, Fundy Trail Parkway). Sometime around 530 million years ago, Avalonia found itself surrounded by the Rheic Ocean between Ganderia and Gondwana. An abundance of shallow marine sediments accumulated from about 530 to 510 million years ago (site 14, King Square West), a testament to the break-up.

Fundy Trail Parkway (site 18).

Terranes. Here are the terrane names from areas explored in this book, arranged roughly from northwest to southeast as currently positioned. Not part of this list but important to mention is the Meguma terrane, now exposed only in neighbouring Nova Scotia. Its prolonged, transcurrent collision with Avalonia had significant effects in New Brunswick (see Story C).

Elmtree (Ganderia). Fragment of ocean crust from the Tetagouche back-arc basin (site 36, Atlas Park).

Miramichi (Ganderia). Passive margin sequence (sites 29, Hays Falls; 38, Middle Landing).

Annidale (Ganderia). Back-arc basin related to the New River terrane (Route 7).

St. Croix (Ganderia). Passive margin of the New River terrane (Route 1).

New River (Ganderia). Magmatic arcs and related plutons (sites 7, Pea Point, Also Nearby; 9, The Anchorage Park; Routes 1 and 7).

Brookville (Ganderia). Passive margin that later became an active margin (sites 10, Barnaby Head; 12, Dipper Harbour; 15, Rockwood Park).

Caledonia (Avalonia). Magmatic arcs and related plutons below Cambrian sedimentary rocks (sites 14, King Square West; 15, Rockwood Park (FYI); 16, Cape Spencer; 18, Fundy Trail Parkway; 19, Point Wolfe).

Collisions: Belts and Basins

Late Ordovician, Silurian, and Devonian periods

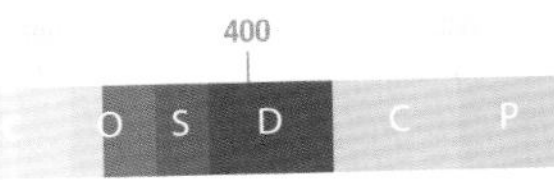

The time period from 470 to 375 million years ago was marked by a long series of continental rearrangements as Ganderia and Avalonia moved toward, and ultimately collided with, Laurentia.

Terrane accretion. Ganderia's leading edge collided with Laurentia to complete the Taconic orogeny around 450 million years ago. As Gondwana continued to approach Laurentia, the Tetagouche back-arc basin closed—its sea floor was subducted under Laurentia. The rest of Ganderia collided with Laurentia in the Salinic orogeny about 430 million years ago.

Todds Point (site 1).

The collision of Ganderia with Laurentia was likely still under way when Avalonia collided with Ganderia around 420 million years ago, causing the Acadian orogeny. In some locations geologists have found it challenging to separate the effects of these two events. By the end of these events, substantially all of the landmass of the present-day region was in place, accreted to Laurentia.

Granite and other igneous rocks. The collision of Avalonia with Ganderia's trailing edge led to widespread igneous activity in Ganderia in the Silurian and Devonian periods. In southern and central New Brunswick, volcanic rocks (sites 32, Williams Falls; 33, Sugarloaf; 34, Inch Arran Point) and igneous intrusions (sites 1, Todds Point; 4, St. George; 28, McAdam; 37, Pabineau Falls) provide evidence of this long period of tectonic unrest.

Kingston belt. This belt of volcanic rocks and intrusions formed as a magmatic arc between 442 and 435 million years ago when a subduction zone formed along the trailing edge of Ganderia (the edge facing away from Laurentia) as Avalonia approached.

Grand Falls (site 30).

Deformation. Both collisions caused rocks to be deformed (for example, sites 30, Grand Falls; 31, Edmundston; 38, Middle Landing), faulted, metamorphosed, and intruded by igneous rocks during the Salinic and Acadian orogenies in the Silurian and Devonian periods. The effects of more distant collisions continued even into the Carboniferous period (see Story C).

Basins. The collisions during this period of time warped the crust downward in some places, forming a variety of marine basins in which sediment accumulated. New Brunswick preserves a record of three such basins.

Matapédia Basin. From 450 to 405 million years ago, sediment accumulated in this basin during the long sequence of plate collisions. Initially associated with a volcanic arc during the Ordovician period, during Silurian and Devonian times it evolved into a foreland basin, where continental crust was warped downward by the weight of an overriding plate (sites 30, Grand Falls; 31, Edmundston).

Fredericton Trough. These rocks (site 26, Mactaquac Dam) formed from sediments deposited 440 to 430 million years ago in the last remains of the Tetagouche back-arc basin (see Story A). As the basin narrowed, sediment derived from both Gondwana and Laurentia was deposited in the trough. The Fredericton fault within the trough marks the site of the basin's ultimate closure during the Salinic orogeny.

Mascarene Basin. These sedimentary and volcanic layers accumulated in a back-arc basin paired with the Kingston belt along Ganderia's trailing edge as Avalonia approached (site 5, Greens Point). In this basin, volcanic and sedimentary layers accumulated from about 450 to 420 million years ago, ending when Avalonia collided with Ganderia. Sedimentary and volcanic rocks from a similar but unnamed basin occur at site 6, Herring Cove.

Assembly of Pangaea: The Maritimes Basin

Devonian, Carboniferous, and Permian periods

In what is now Atlantic Canada, the style of plate interaction changed around 390 million years ago as Meguma and Gondwana moved closer. Instead of mainly convergent motions, more complex transcurrent motions dominated as Gondwana's main landmass approached and collided obliquely with Laurentia to cause the Alleghenian orogeny. Between about 375 and 300 million years ago, a complex system of major strike-slip faults sliced through the region, displacing some blocks of crust by 250 kilometres or more.

Effects of faulting. Evidence of these large fault movements include mylonite (sites 10, Barnaby Head; 16, Cape Spencer); displaced (allochthonous) crustal blocks (site 12, Dipper Harbour); and intensely deformed areas of high metamorphic grade, including the Partridge Island block (site 13, Taylors Island, FYI) and the Pocologan metamorphic suite (Highway Geology, Route 1). In some locations transcurrent movement caused localized volcanic activity (sites 13, Taylors Island; 25, Currie Mountain; 27, Harvey).

Taylors Island (site 13).

Early faulted basins. Initially the transcurrent motions caused only localized depressions. By around 375 million years ago, river sediment was gathering in isolated low-lying areas (sites 2, St. Andrews; 7, Pea Point). Over time as Gondwana's slant-wise approach continued to wrench the crust, the depressions broadened and coalesced into larger basins, eventually encompassing the Maritimes basin as a whole.

Windsor Sea. From about 340 to 325 million years ago, sea level rose and then fluctuated during a time of severe climate change. In extreme heat, small bodies of seawater became isolated and evaporated, leaving gypsum deposits behind (site 22, Hillsborough). After the sea withdrew for the last time, renewed uplift along faults led to rapid erosion, depositing thick wedges of sediment nearby (sites 11, Lepreau Falls; 21, Hopewell Rocks).

River landscape, coal. From about 325 to 305 million years ago, the equatorial climate was tempered by distant polar ice caps. Great rivers flowed across the basin depositing vast quantities of sediment (sites 20, Cape Enrage; 23, Minto; 24, Fredericton Junction; 39, Cap-Bateau; 41, French Fort Cove). Some rivers snaked through lush, swampy forests, the remains of which would turn to coal (site 23, Minto).

Dry climate. Early in the Permian period, the climate in Pangaea's vast, landlocked interior started to become drier. The youngest rock layers from the Maritimes basin are found on Prince Edward Island (sites 42, North Cape; 43, Cavendish Beach; 44, Charlottetown) because it is located near what was the deepest part of the Maritimes basin.

Cavendish Beach (site 43).

Stratigraphic groups. The study of rock layers is known as stratigraphy. As part of this study, geologists give formal names to groups of rock layers to make it easier to talk about them. Each stratigraphic group contains multiple rock formations, and each rock formation comprises a sequence of interrelated rock layers. The formally recognized groups of the Maritimes basin in New Brunswick and Prince Edward Island are, from oldest to youngest:

Horton Group (365–350 million years). Conglomerate, sandstone, and shale deposited in rivers and lakes in early faulted basins: steep, localized valleys along large-scale faults.

Sussex Group (about 350 million years). Similar to the Horton Group, but formed in a later episode of fault movement.

Windsor Group (345–335 million years). Limestone and evaporite formed in the warm, shallow Windsor Sea.

Mabou Group (335–325 million years). Coarse sandstone and conglomerate deposited in alluvial fans accumulating on the former site of the Windsor Sea.

Cumberland Group (about 320 million years). Even-grained, in some places coal-bearing sandstone deposited by large, meandering river systems in a lush, primitive forest.

Pictou Group (315–280 million years). Early coal-bearing but later barren conglomerate, sandstone, and shale deposited by river systems in an increasingly dry climate.

Atlantic: The Fundy Basin

Triassic and early Jurassic periods

From 270 to 200 million years ago, the Earth's landmasses were briefly unified. What is now New Brunswick and Prince Edward Island lay right in the middle of Pangaea, near the equator. But the Earth's system of restless tectonic plates does not favour supercontinents; none have lasted long.

Rift valleys. Beginning about 250 million years ago, a series of broad rift valleys, including the Fundy basin, began to scar Pangaea. The climate in Pangaea's interior was hot and dry, and the basin contained landforms typical of today's desert valleys (site 17, St. Martins).

St. Martins (site 17).

Red Point (site 9, outcrop 1).

Flood basalt. As rifting continued, huge volumes of lava poured through deep fissures in the crust, flooding the Fundy basin with basalt (sites 8 and 9 on Grand Manan Island) as part of a massive igneous event affecting a huge swath across central Pangaea. Related to this, molten rock filled other long, narrow cracks that cut across the region (site 3, Ministers Island).

Many of the early rifts and cracks failed and became inactive. Tectonic activity shifted to a series of interconnecting fractures to the east and south that would become the present-day Atlantic Ocean. By 190 million years ago, Pangaea was no more: Its fragments were drifting apart toward their present-day positions.

Regional Geoheritage

New Brunswick and Prince Edward Island, like the rest of Atlantic Canada, are rich in natural resources. The region's mines and quarries have provided a livelihood and source of pride for generations of residents, just as its forests, farms, and fisheries have. For sites with strong connections to the provinces' geoheritage, check the Trip Planner or look for the geoheritage icon "H" in the site text.

Raw materials. Historically, the rocks of this region have provided raw materials such as oil shale (site 21, Hopewell Rocks), gypsum (site 22, Hillsborough), coal (site 23, Minto), ore (site 38, Middle Landing), and aggregate (sites 25, Currie Mountain; 36, Atlas Park) to industry.

Albert County Museum (site 21, Also Nearby).

Grindstones. The historic grindstone industry relied on fine sandstone of world-renowned quality (sites 20, Cape Enrage; 21, Hopewell Rocks; 39, Cap-Bateau; 41, French Fort Cove).

Building stone. Granite, gabbro, and sandstone of great beauty and permanence grace monuments and buildings both in New Brunswick and Prince Edward Island. The buildings highlighted in this book were built of locally quarried stone and thus display the local bedrock (sites 4, St. George; 21, Hopewell Rocks; 28, McAdam; 31, Edmundston; 37, Pabineau Falls; 40, Grande-Anse to Neguac; 41, French Fort Cove; 43, Cavendish Beach; 44, Charlottetown).

Brass House, Charlottetown (site 44).

Stonehammer Geopark. New Brunswick is home to Canada's first UNESCO Geopark. The geopark features many geological attractions in and around Saint John, and celebrates that city's longstanding connections with the history of geology (sites 11, Lepreau Falls; 13, Taylors Island; 15, Rockwood Park; 17, St. Martins; 18, Fundy Trail Parkway).

Resources

Museums & Interpretive Centres

New Brunswick offers a variety of museums and interpretive centres with geology-themed exhibits, including:

Albert County Museum, 3940 Route 114, Hopewell Cape (www.albertcountymuseum.com/the-land-landing)

Grand Manan Museum, 1141 Route 776, Grand Manan (www.grandmananmuseum.ca)

Minto Museum, 187 Main Street, Minto (www.villageofminto.ca/attractions/minto-museum-and-information-centre)

New Brunswick Museum, Market Square, Saint John (www.nbm-mnb.ca)

Quartermain Earth Science Centre, 2 Bailey Drive, University of New Brunswick, Fredericton (www.quartermainearthsciencecentre.com)

Books

Allaby, Michael. *A Dictionary of Earth Sciences.* Oxford University Press, 2008.

Atlantic Geoscience Society. *The Last Billion Years: A Geological History of the Maritime Provinces of Canada.* Nimbus Publishing, 2001 (reprinted 2019).

Calder, John H. *Island at the Centre of the World: The Geological Heritage of Prince Edward Island.* Acorn Press, 2018.

Canadian Federation of Earth Sciences. *Four Billion Years and Counting: Canada's Geological Heritage.* Nimbus Publishing, 2014.

Fédération canadienne des sciences de la Terre. *Quatre milliards d'années d'histoire: Le patrimoine géologique du Canada.* Nimbus Publishing, 2014.

Martin, Gwen. *For Love of Stone*, vols. 1 & 2 (Publ. No. MSC 8 & MSC 9). New Brunswick Department of Natural Resources and Energy, 1991. Available online at http://dnr-mrn.gnb.ca/ParisWeb.

Martin, Gwen. *Gesner's Dream: Trials and Triumphs of Early Mining in New Brunswick.* Canadian Institute of Mining, Metallurgy & Petroleum, 2003.

Wilson, Reginald A. *The Middle Paleozoic Rocks of Northern and Western New Brunswick, Canada* (Memoir 4). New Brunswick Department of Energy and Resource Development (Geological Surveys Branch), 2017. Available online at http://dnr-mrn.gnb.ca/ParisWeb.

Websites

Maps

Atlas of Canada, atlas.gc.ca/toporama/en/index.html

Natural Resources Canada Geospatial Extraction Tool (free downloadable GIS data), maps.canada.ca/czs/index-en.html

New Brunswick Department of Energy and Resource Development (Minerals and Petroleum). Provincial bedrock maps:

> Visit www2.gnb.ca and search for "Bedrock Mapping" then follow the link to find maps by region. Or visit http://www1.gnb.ca/0078/GeoscienceDatabase and use PARIS Search to find maps by publication name.

Geology

Natural Resources Canada. *WEBLEX: Lexicon of Canadian Geological Names On-line.* http://weblex.nrcan.gc.ca/weblexnet4/weblex_e.aspx. (Search by rock formation name.)

New Brunswick Museum and Virtual Museum of Canada. *Magnificent Rocks*, http://magnificentrocks-rochesmagnifique.ca/

New Brunswick geoscience databases, http://www1.gnb.ca/0078/GeoscienceDatabase

> On this page are links for several provincial databases including publications and maps (PARIS Search), bedrock lexicon database (where you can search by rock formation name), mineral occurrence database, and fossil database.

PEI Museum and Heritage Foundation. *Natural History*, http://www.peimuseum.ca/index.php3?number=1050810

Stonehammer Geopark, https://stonehammergeopark.com

Touring and Hiking

Backroad Mapbooks, www.backroadmapbooks.com

Canadian Hydrographic Service (tide tables), www.tides.gc.ca

Hiking New Brunswick, www.hikingnb.ca

Island Trails (Prince Edward Island), www.islandtrails.ca/en

Leave No Trace, www.leavenotrace.ca

Tourism New Brunswick, www.tourismnewbrunswick.ca

Tourism Prince Edward Island, www.tourismpei.com

These two pages provide an easy way to plan your geological excursions. The trip planner lists the book's 44 sites as they appear in the Table of Contents. Because the book is organized for travel, it's easy to find groups of sites that fit your itinerary, time commitments, and interests. Sites can all be accessed from the provinces' Scenic Drives, as indicated in the list.

For more information about New Brunswick's Scenic Drives, visit the province's tourism website at www.tourismnewbrunswick.ca or consult your copy of the province's current Travel Guide. For Prince Edward Island, visit www.tourismpei.com or refer to your copy of the province's current Visitor's Guide.

No.	Site	Scenic Drive	Route	Hike	Fac.	Rocks	Geo-heritage
TOUR 1 – FUNDY WEST							
1	**Todds Point**	Fundy Coastal Drive	170	1.5	P	i	
2	**St. Andrews**	Fundy Coastal Drive	127	1	P	s	
3	**Ministers Island**	Fundy Coastal Drive	127	3–4	P	i (s)	
4	**St. George**	Fundy Coastal Drive	1	0.7		i	h
5	**Greens Point**	Fundy Coastal Drive	172	0.5	P	i, s	
6	**Herring Cove**	Fundy Coastal Drive	774	0.4	P	i, s	
7	**Pea Point**	Fundy Coastal Drive	176	1.1	P	s	
8	**Southwest Head**	Fundy Coastal Drive	776	0.6		i	
9	**The Anchorage Park**	Fundy Coastal Drive	776	1.5	P	i, m (f)	
10	**Barnaby Head**	Fundy Coastal Drive	175	0.2	P	i (f)	
11	**Lepreau Falls**	Fundy Coastal Drive	175	0.5	P	s	SH
TOUR 2 – FUNDY EAST							
12	**Dipper Harbour**	Fundy Coastal Drive	790	0.2		i	
13	**Taylors Island**	Fundy Coastal Drive	1	1.1	P	i, s	SH
14	**King Square West**	Fundy Coastal Drive	1	0.5	P	s	
15	**Rockwood Park**	Fundy Coastal Drive	1	2.4	P	i, s, m	SH
16	**Cape Spencer**	Fundy Coastal Drive	1	0.1		i (f)	
17	**St. Martins**	Fundy Coastal Drive	111	0.7		s	SH
18	**Fundy Trail Parkway**	Fundy Coastal Drive	111	0	P	i, m	SH
19	**Point Wolfe**	Fundy Coastal Drive	114	1.3	P	i	
20	**Cape Enrage**	Fundy Coastal Drive	915	1	P	s	(h)
21	**Hopewell Rocks**	Fundy Coastal Drive	114	2.3	P	s	(h)
22	**Hillsborough**	Fundy Coastal Drive	114	0.3		s	(h)

Site Name and Number (as found in the Table of Contents and site headings)

Scenic Drive (the provincial scenic route on which the site is located)

Nearest Highway (the highway pictured in the road map for the site)

Hiking Distance (the round-trip distance between the parking location and the main outcrop, in kilometres)

Facilities (P, located in a national, provincial, or community park, or other protected area; M, includes a geology- or geoheritage-themed museum or interpretive centre)

Rock Types (i, igneous; s, sedimentary; m, metamorphic; (f), folding or faulting)

Geoheritage (SH, Stonehammer Geopark site; h, heritage-focussed site; (h), nearby heritage feature(s) described)

No.	Site	Scenic Drive	Route	Hike	Fac.	Rocks	Geo-heritage
TOUR 3 – NORTHWEST							
23	**Minto**	River Valley Scenic Dr.	10	0.5	M	s	h
24	**Fredericton Junction**	River Valley Scenic Dr.	101	0.8	P	s	
25	**Currie Mountain**	River Valley Scenic Dr.	105	2	P	i	
26	**Mactaquac Dam**	River Valley Scenic Dr.	105	0.3		s	
27	**Harvey**	River Valley Scenic Dr.	3	0.8	P	i	
28	**McAdam**	River Valley Scenic Dr.	4	0.5		i	h
29	**Hays Falls**	River Valley Scenic Dr.	165	3.3		s	
30	**Grand Falls**	River Valley Scenic Dr.	108	1–2	P	s (f)	
31	**Edmundston**	River Valley Scenic Dr.	144	0.2	P	s, m (f)	
32	**Williams Falls**	Appalachian Range Rt.	385	0.6	P	i, m	
33	**Sugarloaf**	Appalachian Range Rt.	11	4	P	i	
TOUR 4 – NORTHEAST							
34	**Inch Arran Point**	Appalachian Range Rt.	134	0.2	P	i	
35	**West Point Island**	Acadian Coastal Dr.	134	0.1		i, s	
36	**Atlas Park**	Acadian Coastal Dr.	134	0.2	P	i	
37	**Pabineau Falls**	Acadian Coastal Dr.	430	0.2		i	(h)
38	**Middle Landing**	Acadian Coastal Dr.	360	0.4		m (f)	(h)
39	**Cap-Bateau**	Acadian Coastal Dr.	305	0.2		s	(h)
40	**Grande-Anse to Neguac**	Acadian Coastal Dr.	11	0		s	h
41	**French Fort Cove**	Acadian Coastal Dr.	8, 11	2.5	P	s	h
42	**North Cape**	North Cape Coastal Dr.	12	0.7	P	s	
43	**Cavendish Beach**	Green Gables Shore Dr.	13	0.5	P	s	(h)
44	**Charlottetown**	Charlottetown	1, 2	3.5		s	h

Would you like to learn about or visit rocks from a particular period of this region's long geological history? The book is organized around road travel, but these two pages provide a virtual trip through time, listing attractions in order of age. The "time tour" begins with the provinces' oldest known exposed rocks at site 18, then wends its way through the geologic record, covering a period of about 500 million years.

The ages shown below are based on the best available evidence for each site. For sites with rocks of different ages, the outcrops are listed separately. The ages of many of the igneous rocks have been measured using radiometric methods. The ages of sedimentary rocks are estimates based on fossil evidence and other constraints.

No.	Site	Story	Age (Ma)	Period	Terrane-Basin
18	Fundy Trail Parkway (outcrop 3)	A	690+	Cryogenian	Caledonia
16	Cape Spencer	A	625	Ediacaran	Caledonia
19	Point Wolfe	A	620	Ediacaran	Caledonia
9	Anchorage Park (outcrop 3)	A	618	Ediacaran	New River
9	Anchorage Park (outcrop 2)	A	600	Ediacaran	Brookville
15	Rockwood Park (outcrops 1–2)	A	600	Ediacaran	New River
10	Barnaby Head	A	555	Ediacaran	Brookville
12	Dipper Harbour	A	550	Ediacaran	Brookville
15	Rockwood Park (FYI)	A	550	Ediacaran	Caledonia
18	Fundy Trail Parkway (outcrop 1)	A	550	Ediacaran	Caledonia
15	Rockwood Park (outcrop 3)	A	540	Ediacaran	Brookville
18	Fundy Trail Parkway (outcrop 2)	A	515	Cambrian	Caledonia
29	Hays Falls	A	510	Cambrian	Miramichi
14	King Square West	A	500	Cambrian	Caledonia
38	Middle Landing	B	480	Ordovician	Miramichi
36	Atlas Park	B	460	Ordovician	Elmtree
30	Grand Falls	B	440	Ord/Sil	Matapedia
5	Greens Point	B	437	Silurian	Mascarene
6	Herring Cove	B	430	Silurian	Mascarene
26	Mactaquac Dam	B	427	Silurian	Fredericton
4	St. George	B	425	Silurian	Late Intrusion
1	Todds Point	B	423	Silurian	Late Intrusion
32	Williams Falls	B	420	Silurian	Matapedia
35	West Point Island	B	420	Silurian	Matapedia

Site Number (colour-coded for Tours 1–4)

Site Name (as seen in the Table of Contents and site headings)

Story (as labelled in Regional Background, pp. 18–26)

Age (in millions of years [Ma])

Period (from the geologic timescale; see Geologic Time, pp. 8–9)

Terrane or Basin (as described in Regional Background, pp. 18–26)

No.	Site	Story	Age (Ma)	Period	Terrane-Basin
31	Edmundston	B	415	Devonian	Matapedia
28	McAdam	B	412	Devonian	Late Intrusion
33	Sugarloaf	B	407	Devonian	Matapedia
34	Inch Arran Point	B	407	Devonian	Matapedia
37	Pabineau Falls	B	397	Devonian	Late Intrusion
2	St. Andrews	C	375	Devonian	Maritimes
7	Pea Point	C	375	Devonian	Maritimes
13	Taylors Island	C	360	Dev/Carb	Maritimes
27	Harvey	C	360	Dev/Carb	Maritimes
22	Hillsborough	C	335	Carboniferous	Maritimes
25	Currie Mountain	C	330	Carboniferous	Maritimes
11	Lepreau Falls	C	325	Carboniferous	Maritimes
21	Hopewell Rocks	C	325	Carboniferous	Maritimes
20	Cape Enrage	C	318	Carboniferous	Maritimes
23	Minto	C	310	Carboniferous	Maritimes
24	Fredericton Junction	C	310	Carboniferous	Maritimes
40	Grande-Anse to Neguac	C	305	Carboniferous	Maritimes
41	French Fort Cove	C	305	Carboniferous	Maritimes
39	Cap-Bateau	C	300	Carboniferous	Maritimes
42	North Cape	C	298	Permian	Maritimes
44	Charlottetown	C	287	Permian	Maritimes
43	Cavendish Beach	C	285	Permian	Maritimes
17	St. Martins	D	240	Triassic	Fundy
3	Ministers Island	D	201	Jurassic	Fundy
8	Southwest Head	D	201	Jurassic	Fundy

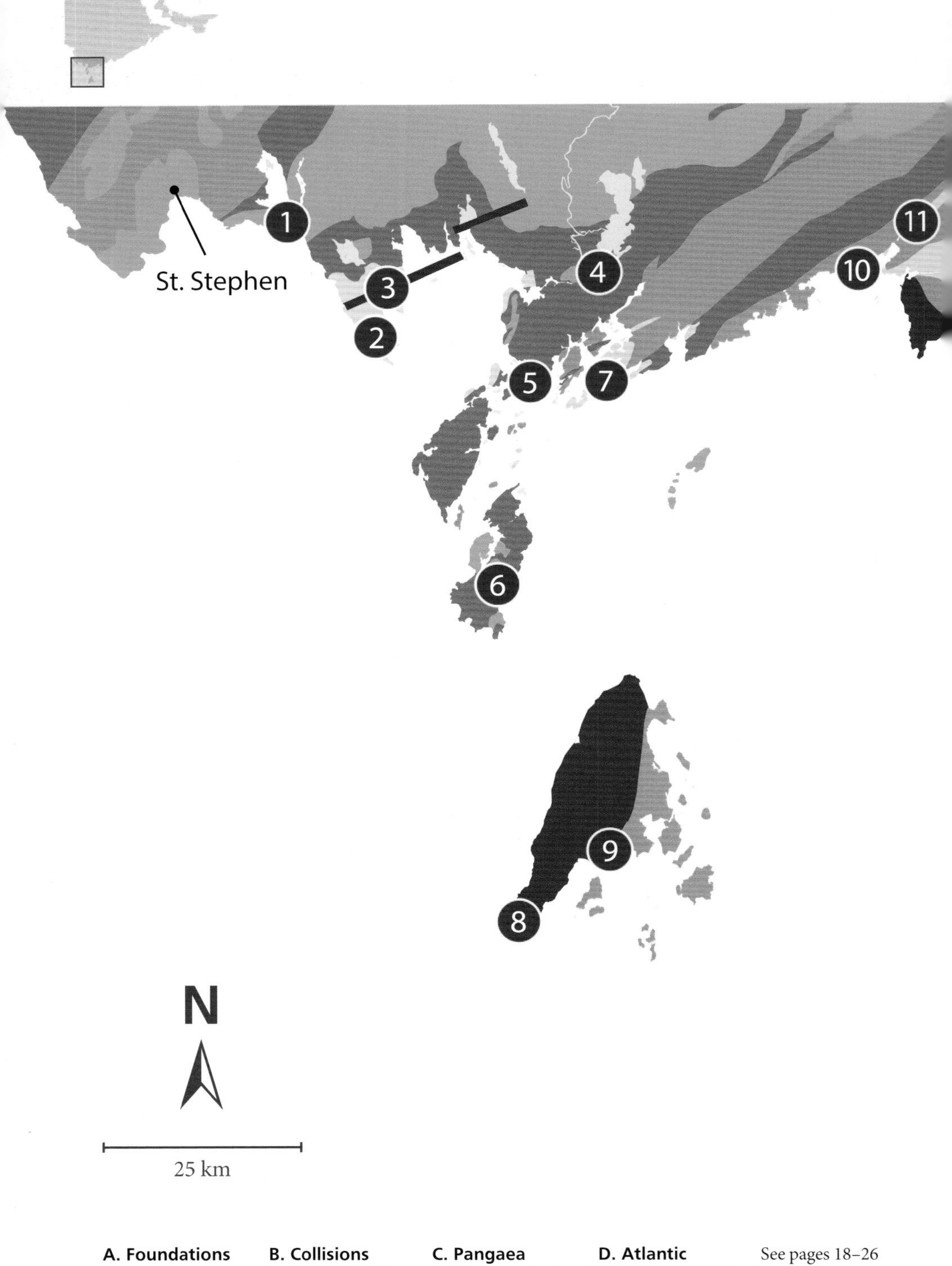

See pages 18–26

FUNDY – WEST

Tour 1 at a Glance

Tour 1 begins near St. Stephen and explores the coast of southwestern New Brunswick, including island destinations accessible by ferry. Ferries and Fundy tides make this a unique way to explore the province's geology and geoheritage. The region includes rocks spanning more than 400 million years of Earth history, including some of New Brunswick's oldest and youngest rocks.

At these sites, you can explore ...

1	**Todds Point**	Magma mingling in a Silurian intrusion
2	**St. Andrews**	The beginnings of the Maritimes basin
3	**Ministers Island**	A gabbro dyke signalling Pangaea's demise
4	**St. George**	Granite monuments from a historic industry
5	**Greens Point**	The complexity of a back-arc basin
6	**Herring Cove**	Rocks formed by turbidity currents
7	**Pea Point**	Colourful conglomerate from a local source
8	**Southwest Head**	Basalt flows of Grand Manan Island
9	**The Anchorage Park**	Ediacaran rocks of eastern Grand Manan Island
10	**Barnaby Head**	A deformed granodiorite of the Brookville terrane
11	**Lepreau Falls**	Sandstone that is probably Carboniferous

Ganong Nature Park at Todds Point offers views of the St. Croix River and historic St. Croix Island.

Champlain's Bubbles

Magma Mingling in a Silurian Intrusion

The year was 1604, as any Canadian schoolchild knows, when a French expedition including explorer Samuel de Champlain sailed up the Bay of Fundy and into the St. Croix River. They chose tiny St. Croix Island as the site of their settlement, enduring one terrible winter before moving across the bay to found Port Royal.

The site of the explorers' first home in North America is just 5 kilometres downriver from Todds Point, and easily visible from Ganong Nature Park grounds on a clear day. To honour Champlain's presence centuries ago, park naturalists gave the name "Champlain's Bubbles" to a tide-related effect that sometimes appears in the water below Ganong Cottage. But the name could equally be applied to nearby rock outcrops.

Along the shore, a process known to geologists as magma mingling is preserved in sharp detail on the wave-washed rocks. Granite and gabbro flowed together without mixing, like oil and water. Bimodal igneous activity—involving both felsic and mafic magma—is common following continental collision. And that's exactly what was going on when the rocky "bubbles" at Todds Point formed.

Getting There

Driving Directions

From Route 170 in St. Stephen, follow Prince William Street eastward, continuing east as the name changes first to Ledge Road and eventually to Todds Point Road.

Or, from Route 1 take Exit 13 to Route 170 and drive east about 1.5 kilometres to Oak Haven Road (N45.22181 W67.19534). Follow Oak Haven Road south for about 6 kilometres to Todds Point Road (N45.17578 W67.20436) and turn left (east).

About 100 metres east of Oak Haven Road, fork right to stay on Todds Point Road. After about 3 kilometres the road ends at the entrance to Ganong Nature Park.

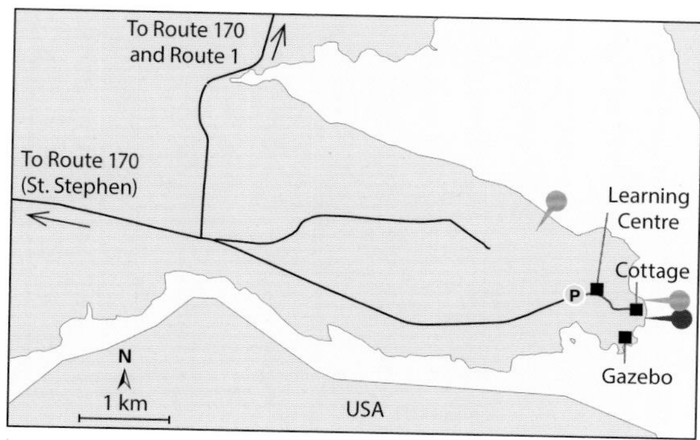

Where to Park

Parking Location: N45.17196 W67.16727

Park in the large gravel area outside the park gate.

Walking Directions

From the parking area, enter through the park gate and follow the gravel track, bearing right around the Quoddy Learning Centre and over the footbridge. The gravel track leads to Ganong Cottage. Bear left around the cottage and follow the Tidal Pool Trail left to a small stairway leading to the beach. From the base of the stairs, turn right and walk south along the shore for about 150 metres to the first outcrop. Continue south about 100 metres for further outcrops.

Notes

For more information about the trails and amenities at Ganong Nature Park, see ganongnaturepark.com.

1:50,000 Map

St. Stephen 021G03

Provincial Scenic Route

Fundy Coastal Drive

On the Outcrop

Rounded areas of dark gabbro are surrounded by pink granite in these excellent examples of magma mingling.

Outcrop Location: N45.17057 W67.16007

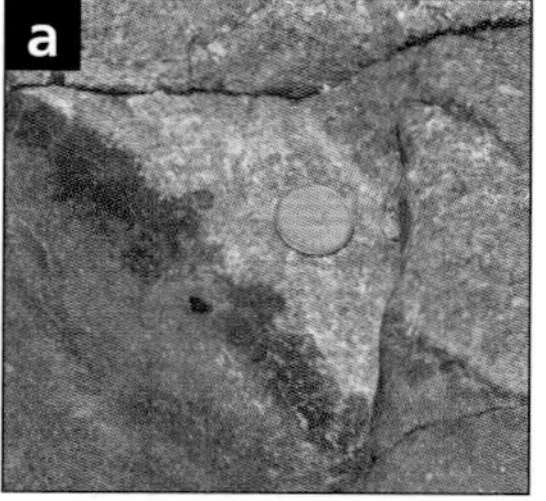

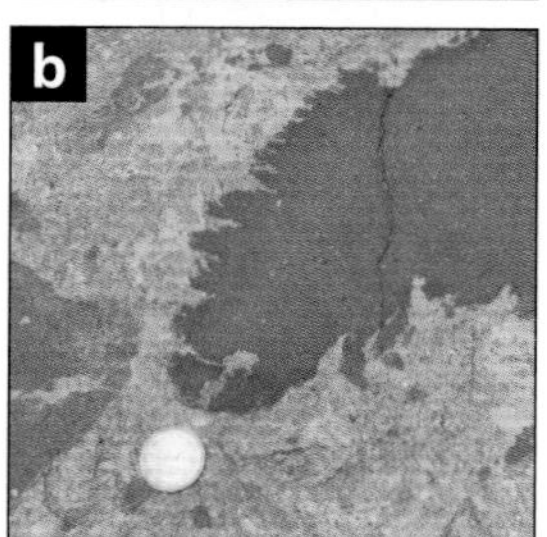

Mingling is such an apt word for the process that formed these rocks. Dark grey gabbro forms large, rounded shapes—the "bubbles"—surrounded by volumes of pink granite. Just like in a retro lava lamp, the two types of magma remained separate as they interacted.

Because molten gabbro is hotter than molten granite, in places the gabbro has a dark, fine-grained chilled margin against the granite (photo **a**). Also characteristic of magma mingling are the wavy or tattered-looking boundaries seen between the two rock types in some outcrops (photo **b**), which could have formed only while both rock types were fluid.

In a few places, straight-sided veins of late-stage granite cut across areas of gabbro that appear to have broken. Because gabbro solidifies at a higher temperature, during cooling it became brittle sooner than the granite magma did.

Rock Unit

Bocabec Complex

Bedrock Map

MP 97-32

FYI

- The gabbro at Todds Point is part of a larger intrusion that extends eastward nearly 20 kilometres and has played a dramatic role in Atlantic history. A small quarry about 14 kilometres east of Todds Point supplied gabbro grave markers for all the victims of the *Titanic* tragedy who were buried in Halifax, Nova Scotia. A team led by geologist Barrie Clarke used rock characteristics and historical records to track down the source.

(a) Grave markers for *Titanic* victims in Fairview Lawn Cemetery, Halifax, Nova Scotia; **(b)** igneous mineral texture in a typical *Titanic* marker.

Also Nearby

At least two additional kinds of granitic rock are found in Ganong Nature Park at Todds Point, as shown below. (**a**) Dark pink granite can be seen on the beach by the Tidal Pool. (**b**) Grey granite with large, visible crystals caps the hill at the top of the park's Lookout Trail.

Exploring Further

Clarke, Barrie, and others. "Forensic Igneous Petrology: Locating the Source Quarry for the 'Black Granite' *Titanic* Headstones in Halifax, Nova Scotia, Canada." *Atlantic Geology*, vol. 53 (2017), pp. 87–114. https://doi.org/10.4138/atlgeol.2017.004.

Beside the St. Andrews Blockhouse National Historic Site of Canada, tilted sandstone layers emerge from a long sandbar at low tide.

Early River Sands

The Beginnings of the Maritimes Basin

St. Andrews has been a popular tourist destination for more than a century, and part of its appeal lies in its well-preserved historic landmarks. St. Andrews Blockhouse is a relic of the War of 1812. One of few such buildings to survive intact, it was erected as a defence against naval incursions along the St. Croix River.

The little park surrounding the blockhouse provides pleasant and ready access to a long, inviting sandbar at low tide. Anchoring the bar, a series of red sandstone outcrops reach up toward the northwest. They are a sort of landmark, too, being among the oldest river deposits to be preserved in the region following a long history dominated by oceans, volcanoes, and constant upheaval.

The sediment in these rock layers accumulated in a basin of limited extent, probably formed by faulting within a mountainous region. Streams readily eroded the high ground, carrying broken rock and sand in a rush down steep slopes in a semi-arid climate. Similar settings exist today, for example, in Turkey or Iran, caught in the collision between Africa and Eurasia.

Getting There

Driving Directions

From Route 1, take Exit 25 or Exit 39 and follow Route 127 to St. Andrews. Where Route 127 reverses direction (N45.08418 W67.06267), continue southeast on Reed Avenue. Turn right (N45.08004 W67.05838) at Harriet Street and turn right again (N45.07723 W67.06070) at Joes Point Road. You may see signs for the St. Andrews Blockhouse National Historic Site. The parking entrance is just past the blockhouse on the left.

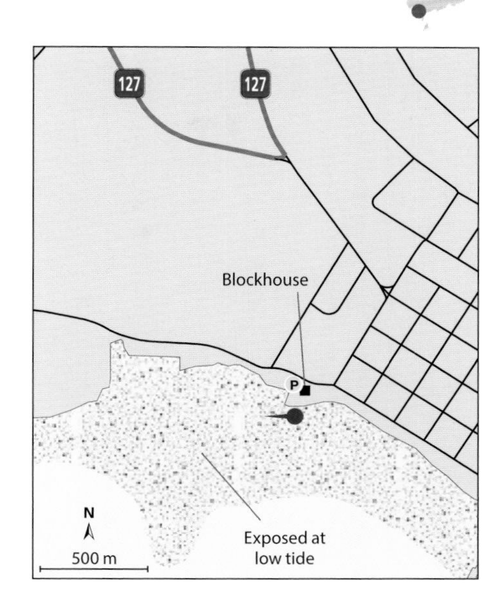

Where to Park

Parking Location: N45.07726 W67.06234

The gravel parking area for St. Andrews Blockhouse is on the west side of the blockhouse grounds.

Walking Directions

A gravel path leads from the parking lot toward the blockhouse, then forks right along the shore. Follow the path to a pink granite stairway and, as conditions allow in low tide, descend to the sandbar, where outcrops are exposed. In lowest tides, clean outcrops extend about 400 metres from shore.

Notes

There is no cost to park or enter the park grounds. For information about National Historic Site tours, check http://www.pc.gc.ca.

Caution

Fundy tides rise quickly. Check tide tables for the day of your visit and plan carefully.

1:50,000 Map

St. Stephen 021G03

Provincial Scenic Route

Fundy Coastal Drive

On the Outcrop

Red, pebbly sandstone cross-beds attest to the river origins of the sandstone layers.

Outcrop Location: N45.07630 W67.06353

Most of the outcrops in the first 200 metres along the sandbar are clear of seaweed, with easily visible features. Because of the way the layers are tilted, their age increases from northeast to southwest across the bar.

The view while facing southeast (toward the pier) cuts across the layers, exposing features characteristic of river sediments. Erosion by the swirling tides has emphasized details like cross-bedding (above) and the wide variety of pebbles mixed into the sediment (detail).

Pebbles ranging from less than 5 millimetres to more than 10 centimetres across are jumbled in a matrix of sand. Fragments of brick-red rhyolite and milky white quartz are common. Their range of sizes and their irregular shapes are both typical of immature sediment. The fragments did not travel far and so did not have time to be worn smooth or broken down into sand.

Immature sediment.

Rock Unit

Perry Formation, Horton Group

Bedrock Map

MP 2005-28

FYI

- By studying the orientation of cross-bedding and other structures in the sandstone, geologists can tell that the river system in this part of the basin flowed from what is now the northwest. The pebbles in the rock closely match the igneous and metamorphic rock types found in that direction.
- Cross-bedding forms as water currents pass over a low place in a sandy surface, for example, beside a ripple, sandbar, or channel. Sand grains generally roll or bounce along with the current, but they fall down any slope they encounter and accumulate there. Over time, the location of the slope moves in the direction of the current as cross-beds build up along the edge.

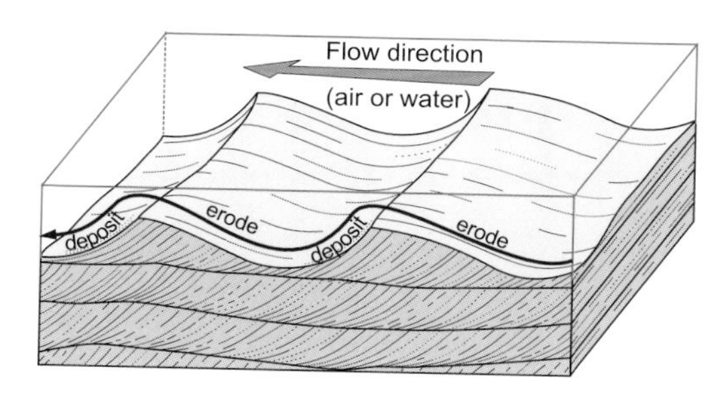

- The stone used to build the retaining wall and stairs at this site is a variety of granite from St. George (site 4).

Also Nearby

The sandstone on Ministers Island (site 3) is part of the same rock formation, deposited in the same basin at around the same time.

Red sandstone layers form low cliffs along the shore on Ministers Island.

Well-maintained trails through historic grounds, meadows, and woodland provide easy, scenic access to features of interest on Ministers Island.

Breaking Up

A Gabbro Dyke Signalling Pangaea's Demise

This island made the perfect retreat for an eccentric millionaire. It's connected to the mainland by a sandy causeway for five hours during each low tide but surrounded by a moatlike channel of deep, chilly seawater the rest of the time. Anyone distracted by the island's charms or miscalculating the tides must implement Plan B—find a boat or stay put while Fundy waters rise and fall.

Once the site of a hobby farm and the summer home of Canadian Pacific magnate and polymath William Cornelius Van Horne, Ministers Island is now a heritage site (with museum) and nature preserve. It's a perfect destination for history buffs, birdwatchers, hikers—and rock enthusiasts.

For the latter, the island offers a rare glimpse of an elusive giant, part of a 700-kilometre-long dyke system that extends into New England and may have fed great lava flows like those on Grand Manan Island (site 8, Southwest Head). It formed at a crucial turning point as Pangaea began to rift apart about 200 million years ago. The Ministers Island dyke, as it is known, was one small step toward the opening of the Atlantic Ocean.

Getting There

Driving Directions

From the eastern branch of Route 127 just outside St. Andrews, turn east (N45.08920 W67.06503) onto Bar Road and follow it for about 1.4 kilometres to the shore. In low-tide conditions, you can drive your vehicle across the gravel bar, which extends about 750 metres from the mainland to Ministers Island. Stop at the kiosk if it is staffed.

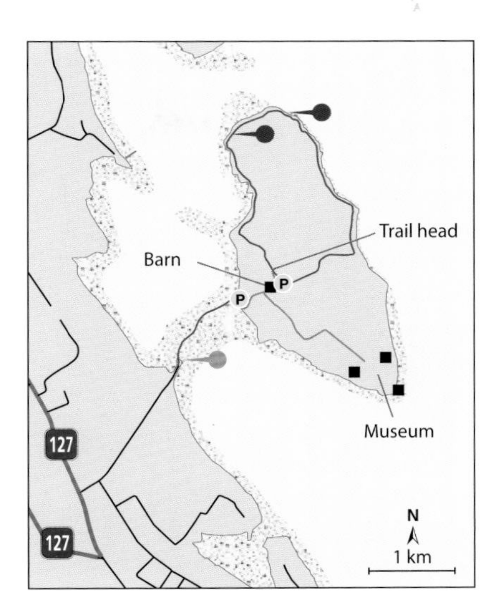

Where to Park

Parking Location: N45.10544 W67.04428

From the kiosk at the shore, follow the gravel drive up the slope. Continue straight up the hill until you just pass the barn, and park in the grassy parking area beside it. (For free parking, leave your car at the shore.)

Walking Directions

Walk north along the gravel track beside the barn, then continue north across the mowed grass. Follow the Perimeter Trail (markers P1, P2, etc.) north along the island's western shore. About 250 metres beyond marker P3, watch for a short gravel section in a low area on the trail (N45.11662 W67.04891). As conditions allow, cross a short distance through the woods to the shore. In this cove is the western outcrop. From there, walk along the shore around the north end of the island to the eastern outcrop location.

Notes

This map shows only selected trails on the northern half of the island. For a detailed map and information about other features and amenities, visit http://www.ministersisland.net.

Caution

Fundy tides rise quickly, and at high tide the sandbar is deep under water. Consult tide tables for the day of your visit and plan carefully. You must leave the island in time to cross the sandbar again.

1:50,000 Map

St. Stephen 021G03

Provincial Scenic Route

Fundy Coastal Drive

On the Outcrop

Dark grey, vertically fractured rock of the Ministers Island dyke is conspicuous between horizontal sandstone layers near the north end of the island.

Outcrop Location: N45.11820 W67.04338

As you round the shoulder of the shoreline into this little cove, the contrast in rock types is sure to catch your eye. The dyke's dark grey gabbro has broken into narrow vertical slabs, the edges of which have been slightly rounded by the action of the surf at high tide. The dyke cuts very neatly through nearly horizontal beds of sandstone, which was hardened and discoloured by contact with the hot magma and forms a sharp-edged buttress against the elements. Because of the excellent wave-washed exposure, you can find the actual contact surface between the two rock types—a rare treat in geology.

Wave action keeps the rock surfaces clean. Some are speckled with visible light and dark minerals; others are strongly pitted, as easily dissolved iron-rich mineral grains have been removed by weathering. Horizontal banding visible on some vertical edges are not rock features; rather, they are due to staining by water and algae.

Gabbro.

Rock Unit

Ministers Island Dyke

Bedrock Map

MP 2005-28

FYI

- If you are pressed for time, the west side provides easy access to a smaller outcrop of the dyke (N45.11660 W67.04925; see Getting There). Look for small vertical slabs of the dyke rock emerging from the bank, as seen on the left side of the photo below.

The Ministers Island dyke (left) and adjacent sandstone (right) on the west side of the island.

- The basalt flows of Grand Manan Island (site 8, Southwest Head) are closely related in age and composition to this and other dykes in the region—all formed about 200 million years ago.

- The Ministers Island dyke (MI) is one of several in the region. Others include the Caraquet and Lepreau River dykes (CD, LR) of New Brunswick, the Christmas Cove dyke (CC) of Maine, and the Shelburne dyke (SD) of Nova Scotia.

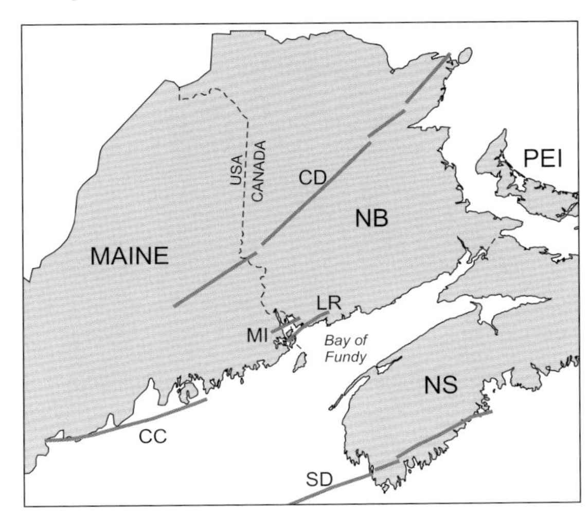

Also Nearby

Outcrops of dark grey igneous rock are exposed elsewhere on Ministers Island and also on the mainland near the island causeway. These volcanic rocks are much older than the dyke. They formed about 375 million years ago, at the same time as the red sandstone on the island and at St. Andrews (site 2).

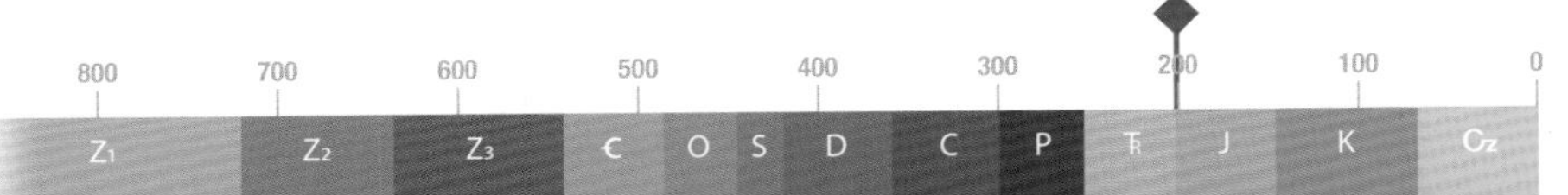

The granite-clad Canada Post office at the west end of Main Street is a fitting landmark in St. George, nicknamed "Granite Town."

Granite Town

Granite Monuments from a Historic Industry

St. George is a venerable town, founded in 1784. In years past various industries including sawmills and shipyards fuelled the town's economy. But local stone quarries produced the most widespread and lasting impact, operating from 1872 to 1953. Their most sought-after product was "St. George Red," a stone of rich, fiery colour, often highly polished. It stole the hearts of architects as far away as New York, Boston, and Ottawa.

Easily accessed natural outcrops of New Brunswick granite are rare due to the effects of climate and vegetation. A geologist who mapped this area once described a landscape underlain by granite as "dismal ... a wasteland of swamps." Abandoned quarries exist locally, but such settings are notoriously hazardous.

Fortunately, for those open to the idea, St. George's historic cemeteries are showcases for the pink and red local granite, as well as for some local gabbro in more sombre tones. Formed during tectonic unrest subsequent to the collision of Avalonia and Ganderia, these igneous rocks have been rendered into symbols of peaceful rest by the skill of local craftsmen.

Getting There

Driving Directions

(**1**) From Route 770 (L'Etete Road) east of St. George, watch for the intersection with Carleton Street (N45.12911 W66.81915) and turn east. You may see a sign for St. George Rural Cemetery. Follow Carleton Street into the cemetery grounds. (**2**) Return to Route 770 and travel north about 300 metres to its intersection with Main Street (N45.13196 W66.82018). Turn west (left) and follow Main Street to any convenient parking location.

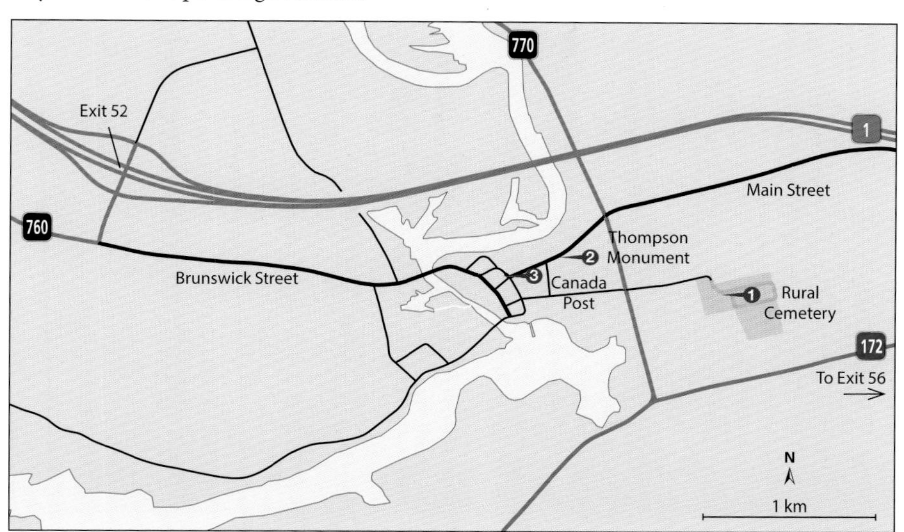

Where to Park

Parking Location: N45.12905 W66.81433

At the cemetery, park beside the open grassy area on the high ground near the entrance. In the St. George town centre, use any convenient street or lot parking space, positioning yourself for walking along Main Street between Clinch and Portage Streets.

Walking Directions

In the cemetery, follow established lanes to visit the historic granite and gabbro monuments in the western end. In town, follow Main Street to the historic Loyalist Cemetery behind St. Mark's Anglican Church. Then walk to the west end of Main Street to see the Canada Post building and granite memorial monument.

1:50,000 Map

St. George 021G02

Provincial Scenic Route

Fundy Coastal Drive

On the Outcrop

The historic section of St. George Rural Cemetery is a showcase for the finely crafted monuments once produced in the area.

Outcrop Location: N45.12892 W66.81404

As you enter the cemetery and follow the paved lane to the right and up a rise, on the left beside a large tree is the Johnson monument (photo **a**), a tall column in "St. George Red," known to geologists as the Utopia granite (see detail). The monument clearly illustrates the glowing colour, high polish, and precise craftsmanship that made the St. George quarries so renowned.

If you stand with your back to that monument and look across the grass to the skyline, the hills you see are those of the granite itself. The old quarry sites are only 4 to 5 kilometres away (see FYI).

On the slopes below the Johnson monument, grassy lanes run among other historic monuments in the western end of the cemetery (photo **b**). The colour palette—bright to pale pink, or black to dark grey—represents the bimodal, granite-gabbro intrusions typical of the region.

Utopia granite.

Rock Unit

Utopia Granite

Bedrock Map

MP 2005-27

FYI

- If you don't want to visit a cemetery that is still in use, consider viewing the Thompson Monument (**a**, below) on Main Street beside St. Mark's Anglican Church (N45.13036 W66.82254). It was the first to be produced of local granite, around 1874. The adjacent Loyalist Cemetery also offers examples of local stone and stone craft.

- For a civic display of local stone, the Canada Post building (**b**, above; N45.12961 W66.82548) at the end of Main Street is partly clad in local granite (the grey stone is from elsewhere). In front of the building the town has erected a monument of St. George red granite.
- At the height of the industry, St. George hosted dozens of quarries just north of the town. Locations of the more prominent operations are shown in the map at right, in which the Saint George batholith appears in pink. The town itself (yellow dot) lies on sedimentary rocks of the Mascarene basin (shown in grey).

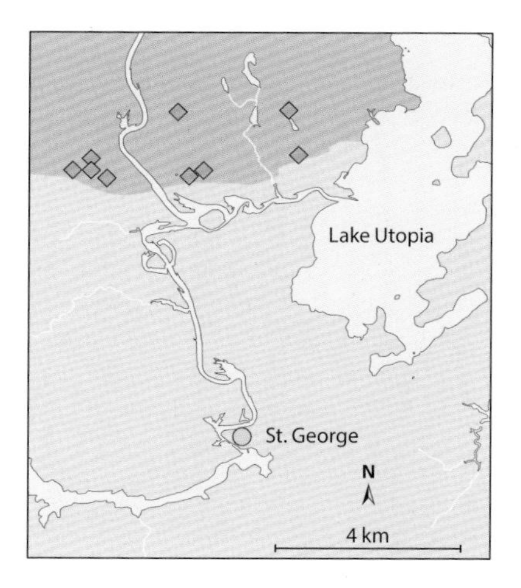

Also Nearby

You may see pink or red granite from this area in a variety of nearby locations, mainly as armour stone; for example, St. Andrews blockhouse (site 2), the end of Bar Road on the approach to Ministers Island (site 3), and the wharf at Letete (near site 5, Greens Point).

Exploring Further

Martin, Gwen. *The Granite Industry of Southwestern New Brunswick: A Historical Perspective.* Fredericton: New Brunswick Department of Energy and Mines, 2013. https://www1.gnb.ca/0078/geosciencedatabase (use PARIS Search, category Popular Geology Paper).

Greens Point lighthouse overlooks Letete Passage and its many islands.

Water, Fire

The Complexity of a Back-Arc Basin

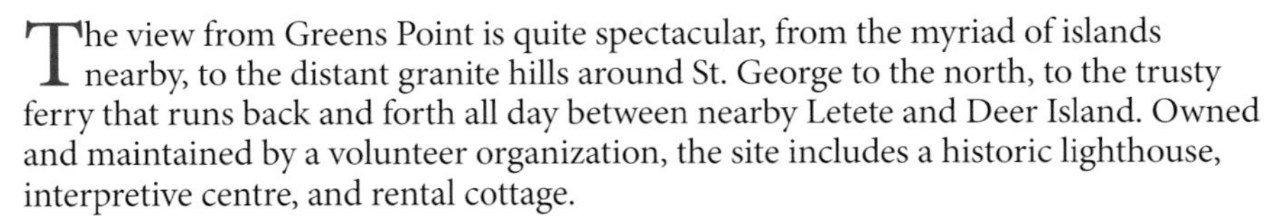

The view from Greens Point is quite spectacular, from the myriad of islands nearby, to the distant granite hills around St. George to the north, to the trusty ferry that runs back and forth all day between nearby Letete and Deer Island. Owned and maintained by a volunteer organization, the site includes a historic lighthouse, interpretive centre, and rental cottage.

It's also host to a dizzying array of rock types, all part of the Mascarene basin. A busy place throughout the Silurian period, this marine basin formed beside a volcanic arc. In consequence it was a repository for all sorts of eruptions, from lava flows to fiery effusions of ash to explosive clouds laden with volcanic debris. In quieter periods, sediments were rapidly eroded from the emerging land and accumulated in layers, only to be covered during the next eruption. All this went on for 20 million years or more.

Back-arc basins are a geological paradox. They open above a subduction zone—an area that is, on a larger scale, closing to become a site of continental collision. The deformed layers on view here were caught up in such a fate.

Getting There

Driving Directions

Follow Route 172 toward the Deer Island ferry terminal. About 350 metres east of the ferry terminal, instead of following Route 172 to the right, fork left (N45.05180 W66.89047) onto Greens Point Road. Follow Greens Point Road for about 1.7 kilometres. After crossing two short causeways, the road ends at the Greens Point Lighthouse grounds.

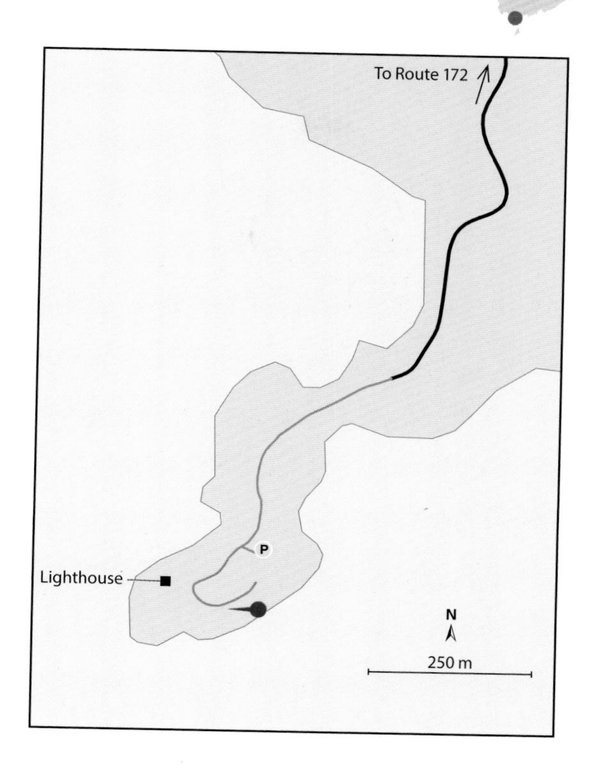

Where to Park

Parking Location: N45.03915 W66.89052

This location is the lighthouse parking area. If the lighthouse gate is closed when you arrive, park beside the road just outside the gate.

Walking Directions

From the parking area, follow the lane toward the lighthouse and then around the point. Near the concrete helicopter landing pad, as conditions permit, cross the grass to view the rock exposures along the shore.

Notes

Greens Point Road is paved but narrow. Wider vehicles should be prepared to yield for oncoming traffic. The Deer Island ferry is just minutes away and runs all day throughout the year.

1:50,000 Map

St. George 021G02

Provincial Scenic Route

Fundy Coastal Drive

On the Outcrop

Once-horizontal layers of volcanic rock stand nearly vertical around the shores of Greens Point.

Outcrop Location: N45.03855 W66.89100

As you walk around the lighthouse grounds, you'll find several places where you can view clean outcrops of rock from the safety of the path or adjacent grass. The majority of these outcrops are made of rocks known as tuff, formed during pyroclastic ("fiery fragment") eruptions that threw hot volcanic debris into the air. Some material fell straight into the waters of the Mascarene basin; some fell onto the volcanic landscape and was later washed into the basin as sediment.

Crystal-lithic tuff.

Subtle colour variations from mauve to greenish grey reflect a range of rock compositions from felsic to mafic. The nearly vertical layers preserve many volcanic features despite their deformation. On a rock pavement near the helicopter pad, look for the fine layering of the volcanic ash, mingled with larger, irregular fragments. They look just like what they are: blobs of lava that hardened as they fell.

Rock Unit

Letete Formation, Mascarene Group

Bedrock Map

MP 2005-52

FYI

- Bright, white pods of quartz visible at Greens Point formed during metamorphism as the Mascarene basin was caught up in the collision of Avalonia with Ganderia. Metamorphic fluids extracted silica from sedimentary and felsic volcanic layers and later deposited it.

Also Nearby

Take the ferry (it's free) from Letete to Deer Island for more chances to explore rocks of the Mascarene basin. During low tide, on the little beach by the ferry terminal on Deer Island (N45.02359 W66.93855), near the turf look for a smoothed, greenish grey rock speckled with white—a vesicular basalt of the Mascarene basin. Next to it is a rust-stained black, layered rock that is being eroded—a black shale. This rock originated as an organic-rich mud on the floor of the Mascarene basin.

Black shale near Deer Island ferry terminal.

At Deer Island Point Park (N44.92589 W66.98437), layered volcanic rocks of the Mascarene basin frame the beach at either end. If you look closely along the eastern end of the beach (shown in photo), in addition to greenish grey volcanic rocks you may see a small gabbro dyke. Identify it by its smoother surface and pattern of easily visible, interlocking black and white crystals.

Volcanic rock in Deer Island Point Park.

Exploring Further

For information about ferry service to Deer Island, visit http://www.eastcoastferriesltd.com.

800 700 600 500 400 300 200 100 0

Z1 Z2 Z3 Є O S D C P Ŧ J K Cz

A viewing platform and picnic bench provide scenic views of the beach and of the rock outcrop below it at Herring Cove, Campobello Island.

Whooshing Down

Rocks Formed by Turbidity Currents

American president Franklin D. Roosevelt and his family made their summer home on Campobello Island, which is easily accessed from the coast of Maine. Now the Roosevelt Campobello International Park commemorates those visits with a museum and nature preserve. Adjacent to it is New Brunswick's Herring Cove Provincial Park.

Herring Cove beach is a sandy crescent more than 2 kilometres long. At its northern end, rocks of a Silurian back-arc basin form a small headland that is easily visited from the beach at low tide. The headland is formed of sedimentary layers intruded by a gabbro dyke.

This is a fitting combination of rock types for a tectonically active setting like a back-arc basin. Erosion of volcanoes in the nearby Kingston arc produced great volumes of sandy sediment, which washed into the basin and piled up its sloping sides. Periodically the sediment became unstable—earthquakes were likely frequent—and rushed downslope in a high-speed, underwater landslide known as a turbidity current. The gabbro dyke is thought to have fed eruptions of lava seen between turbidite layers elsewhere on the island.

Getting There

Driving Directions

Land route: From Route 189 in Lubec, Maine, cross the Roosevelt Bridge and international boundary into New Brunswick. Follow Route 774 for about 4.6 kilometres. You'll pass the Roosevelt Campobello Park visitor's centre, then about 500 metres beyond a sharp right turn, fork right (N44.88406 W66.94523) onto Herring Cove Road. Travel about 1.7 kilometres along Herring Cove Road to the parking area by the shore.

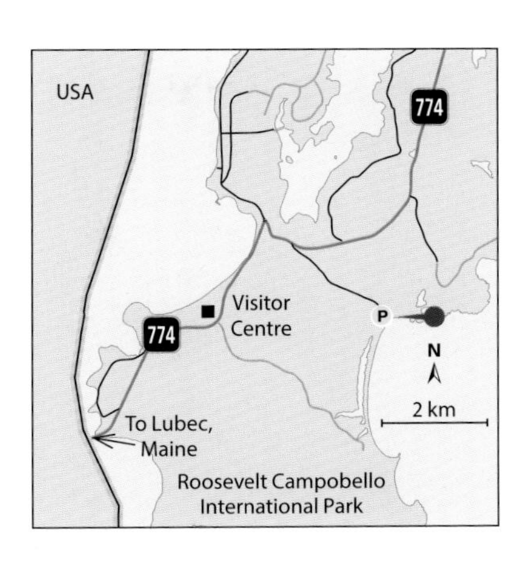

Ferry route: From the ferry terminal in Welshpool, follow North Road south to Route 774. Travel east on Route 774 to Herring Cove Road and continue as described above.

Where to Park

Parking Location: N44.87523 W66.92883

This is a paved parking area near the beach.

Walking Directions

From the southwest corner of the parking lot, follow the gravel trail over a footbridge to the beach. At the beach, in low-tide conditions, walk left (east) toward the rocky bluff. If high water prevents shore access, instead of crossing the bridge, from the southeast corner of the parking lot follow the hiking trail eastward and uphill to the lookout above the bluff.

Notes

The land route to Campobello from Maine crosses an international boundary, requiring passport control, customs, and other formalities. The status of ferry service from Deer Island to Campobello has been uncertain at times. To check on the service, visit http://www.eastcoastferriesltd.com.

Caution

Fundy tides rise quickly. Watch for changing water levels as you explore.

1:50,000 Map

Campobello Island 021B15

Provincial Scenic Route

Fundy Coastal Drive

On the Outcrop

Turbidite layers (orange brown, far right and left) cut by a dyke (greenish grey, centre) form a small headland along the north end of the beach.

Outcrop Location: N44.87470 W66.92775

The headland includes two rock types, made obvious by their contrasting appearance. The turbidite is strongly layered, with the layers tilted up toward the lookout. Because the sediment contains a lot of pyrite, that rock type is stained orange and brown. The gabbro dyke is greenish grey and has no obvious layering. Like a vertical wall, it cuts straight through the turbidite, seeming to emerge right out from under the viewing platform.

In the turbidites, the beds are typically graded (see FYI). Along the edge of the bed, the rock surface feels quite gritty at the base and less so at the top, like different grades of sandpaper. This happened as the turbidity current gradually lost energy and slowed down. The largest grains, being heaviest, were the first to stop moving. Smaller grains remained suspended longer, settling more slowly. Each bed with a coarse base layer represents the start of a new turbidity current.

Turbidite layers.

Rock Unit

Quoddy Formation, Kingston Group

Bedrock Map

MP 2005-25

FYI

- Campobello's combination of Silurian turbidites and dykes is unique in the region, with similarities to both the Kingston belt (a volcanic arc) and the adjacent Mascarene basin (a back-arc basin).
- Turbidity currents are dense, gravity-driven clouds of water, sand, silt, and mud. Sedimentologist Arnold Bouma first realized their significance when he noticed that a distinctive sequence of sedimentary rock layers was common to many deepwater deposits.
- A Bouma sequence begins with a thick layer, or bed, of graded sandstone. The turbidites at this site formed near the source of the sediment. Some sequences here show all five parts, but others are incomplete.

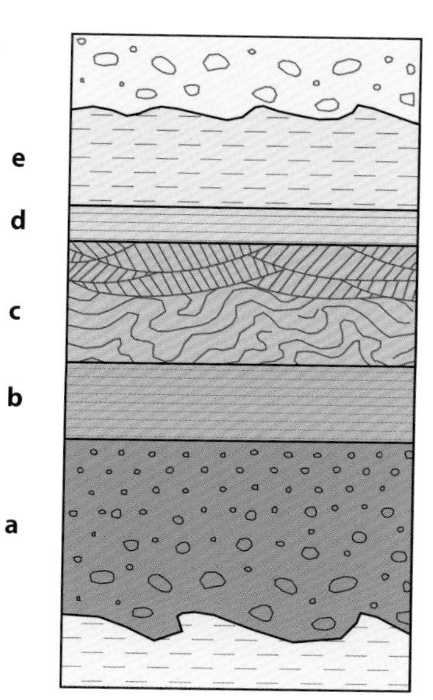

A sketch showing the five parts of an idealized Bouma sequence: (**a**) rapidly deposited, graded sand, sometimes with a gravel base; (**b**) thin, parallel laminations of sand; (**c**) rippled or convoluted layers of silt; (**d**) thin, parallel laminations of silt; and (**e**) mud.

Also Nearby

If you explore northward along the trail that passes the lookout, along the shore you may find outcrops of basalt that erupted underwater onto the surface of the turbidite sands. These may have been fed by the dyke seen at the beach.

The rocks at Greens Point and Deer Island (site 5) formed in a related Silurian back-arc basin, the Mascarene basin.

Exploring Further

Alan, Donald. *A Geological Tour of the Roosevelt Campobello International Park.* Roosevelt Campobello International Park Commission, 2000. https://www.fdr.net/nature.php (bottom of page, under Ecology).

Glacially smoothed outcrops of conglomerate emerge from the beach in the Connors Brothers Nature Preserve at Pea Point.

Pudding Stone

Colourful Conglomerate from a Local Source

If you find yourself in Blacks Harbour (perhaps on the way to Grand Manan Island, sites 8 and 9), consider setting aside time for an easy stroll through the nature preserve at Pea Point. A beach at the end of the preserve's wooded lane offers ready access to the Fundy shore and in normal conditions is accessible even at high tide.

The reward for your efforts is a conglomerate containing large, colourful stones of ancient origin. The outcrops also preserve surface features formed during the last ice age. On the glacially smoothed exposures, it's easy to see details of the pebbles, cobbles, and boulders embedded in this rock, sometimes known as pudding stone.

Geologists recognize that most of the stones (known as clasts) in the conglomerate come from the nearby, ancient New River terrane, about 650 to 550 million years old. Late in the Devonian period the newly assembled landmasses of Ganderia and Avalonia were sliced by large-scale faults. It was beside one such fault that rivers deposited the wedge of rock and sand that later became the Pea Point conglomerate.

Getting There

Driving Directions

From Exit 60 on Route 1, follow Route 176 south about 10 kilometres toward the Grand Manan ferry terminal. Along the way, you'll pass through the town centre of Blacks Harbour. Then about 400 metres from the ferry terminal the road curves right. Look for a gravel pull-off on the left. Turn in to park at the trail head for the Connors Bros. Nature Preserve at Pea Point.

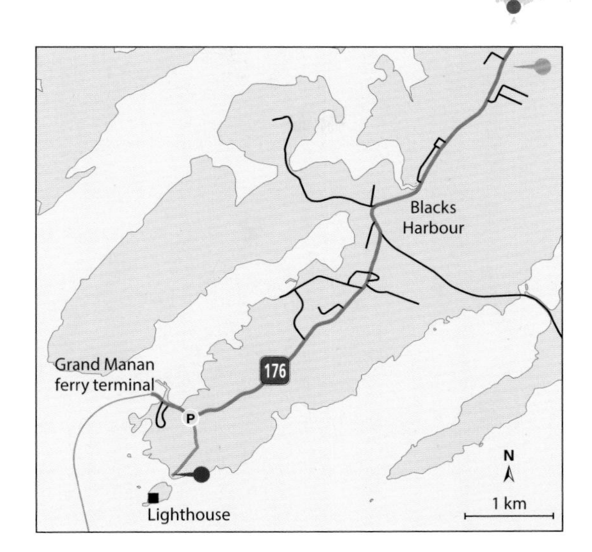

Where to Park

Parking Location: N45.04507 W66.80427

Park in the large gravel pull-off at the trail head on the south side of Route 176.

Walking Directions

From the parking area, walk through the park gate and follow the gravel lane for about 550 metres through a wooded area. The lane ends at the shore. Walk about 10 metres across the sand and pebble beach onto the rounded outcrops.

Notes

Under favourable conditions, it is possible to visit Pea Island and its lighthouse during low tide. However, plenty of clean outcrops are visible on the mainland shore at Pea Point, even in most high tides.

Caution

Fundy tides rise quickly. Check tide tables for the day of your visit and plan carefully if you choose to visit Pea Island.

1:50,000 Map

St. George 021G02

Provincial Scenic Route

Fundy Coastal Drive

On the Outcrop

Glacial action sculpted the beach outcrops into smooth, asymmetrical forms that make it easy to observe the assortment of rounded stones in the conglomerate.

Outcrop Location: N45.04088 W66.80619

As you walk onto the beach, the effects of glaciation may draw your attention first. The rock outcrops are asymmetrical, with shallow slopes on the west side and steep, broken edges on the east side. These fancifully named *roches moutonées* ("rocks like sheep") formed as ice scraped the landscape from west to east. You may also notice parallel, millimetre-scale scratches (known as striations) made by rock debris in the glacial ice. Larger rocks embedded in the ice formed larger, wavelike grooves.

The two most abundant rock types in the conglomerate are easily recognizable. Most of the speckled, grey clasts are from the venerable Blacks Harbour Granite (see Also Nearby). In shades of pink and red are types of granite and rhyolite that geologists recognize as characteristic of the nearby New River terrane. Clasts of all sizes are jumbled together, suggesting that they were deposited rapidly, as would be expected along a steep fault scarp.

Clasts in conglomerate.

Rock Unit

Perry Formation, Horton Group

Bedrock Map

MP 2005-51

FYI

- Conglomerates like those at Pea Point typically accumulate in structures called half-grabens. A half-graben forms when a block of crust becomes tilted as one side slides downward along a fault (see diagram).

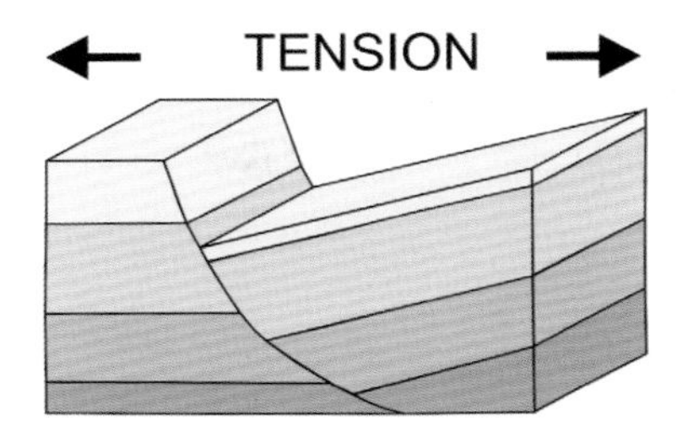

- In this case, water rushed and tumbled down the steep side of the half-graben, quickly depositing the rock debris near its source. Because the clasts travelled only a short distance, they had no chance to break into smaller pieces.
- Although conglomerate forms less than 1 per cent of all sedimentary rocks, it is popular among geologists because it provides plentiful evidence for the source of the eroded material. And because it indicates powerfully flowing water, conglomerate can help geologists envision the ancient landscape in which it formed.

Also Nearby

The Blacks Harbour Granite is hidden beneath the sedimentary rocks at Pea Point but is conspicuous at the local baseball park on the north end of Blacks Harbour (Dr. McLean Memorial Field). About 1.5 kilometres north of Blacks Harbour town centre on Route 176 (Main Street), watch for the outcrop on the right (N45.07121 W66.77141; see photo).

The granite is about 620 million years old (not 1.5 to 1 billion as claimed on the sign at Pea Point trail head). Its surface is discoloured by weathering, but its igneous texture is still apparent.

The lighthouse at Southwest Head stands on high cliffs of basalt on Grand Manan Island in the Bay of Fundy.

Dark Cliffs

Basalt Flows of Grand Manan Island

Of all the Fundy isles, Grand Manan Island is the largest and the most remote. And thanks to its geology, it's also the highest, with basalt cliffs reaching 125 metres. Even on a sunny day, the looming rocks' dark colour feels dramatic. If you sensed something big happened here, you'd be right.

At the end of the Triassic period, large faults opened in many parts of the Pangaean supercontinent as it began to rift apart. As part of that process, huge volumes of red-hot lava flooded out into a valley now known as the Fundy basin. Some eruptions spread over the ground in thick layers. The tall cliffs at Southwest Head formed within a deep lake of lava that accumulated during one of these events.

Initially about 1200°C when molten, the lava solidified and shrank slightly as it cooled. The gradual drop in temperature and the uniform texture of the basalt caused a regular network of cracks to form, initially on the surface. The cracks grew downward through the rock as it cooled, eventually forming the rock columns preserved here.

Getting There

Driving Directions

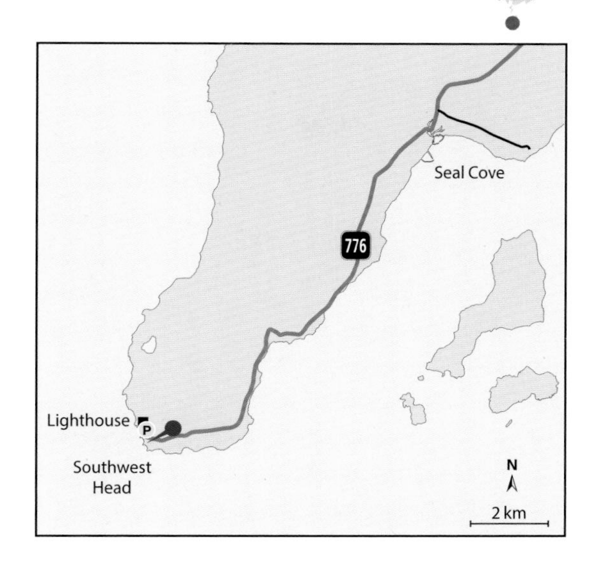

Grand Manan Island is accessible by ferry from Blacks Harbour on the mainland Fundy shore. On Grand Manan Island follow Route 776 south. About 10 kilometres southwest of Seal Cove, Route 776 ends near the lighthouse on Southwest Head.

Where to Park

Parking Location: N44.60060 W66.90515

Park in the unpaved lot beside the lighthouse.

Walking Directions

From the parking location, explore the trails that extend north and south along the clifftop from the lighthouse. As conditions allow, the trails offer excellent views of the basalt cliffs. The footpath north of the lighthouse begins near the buildings; follow it rather than the wider vehicle track, since the footpath provides better views of the rock cliffs.

Caution

Do not approach or stand on the cliff edge. Basalt columns are inherently fractured and may collapse.

Notes

For more information about the hiking trails and lighthouses of Grand Manan and neighbouring islands, visit http://www.grandmanannb.com/hiking.html.

1:50,000 Map

Grand Manan Island 021B10

Provincial Scenic Route

Fundy Coastal Drive

On the Outcrop

Columns of basalt, weathered brown, form a high cliff at Southwest Head.

Outcrop Location: N44.59949 W66.90513

Don't rely on the GPS waypoint to find a safe place to view the cliffs. Evaluate current conditions carefully.

Polygonal columns.

Decades ago in Hawaii, geologists watched and measured a cooling lava lake, staying all day and night sometimes. Minutes after the lava stopped flowing, a honeycomb-like pattern of cracks formed on the surface. For months as the lava solidified and continued to cool, the cracks spread downward, creating tiny earthquakes and audible sounds, mostly during cooler hours in the dark of night.

At Southwest Head, your imagination will have to provide the sound effects, but the process was the same. This type of rock formation is known as columnar basalt. The orientation of the columns records the direction from which the rock cooled. Slow cooling leads to larger, more regular columns, while faster cooling leads to irregular, small columns.

Rock Unit	**Bedrock Map**
Dark Harbour Basalt	MP 2011-14

FYI

- Flood basalts like those on Grand Manan Island affected a large area of Pangaea as it began to rift about 200 million years ago. Including related intrusions, about 10 million square kilometres were affected by this igneous activity. The Central Atlantic Magmatic Province, as it is known (shown in purple in the map at right), was dispersed by the opening of the Atlantic Ocean. Evidence of it is now found in North and South America, Africa, and Europe.
- Rocks of the Ministers Island dyke and related intrusions (site 3, FYI) are similar in chemical composition to the basalt flows of Grand Manan Island and appear to have fed magma to the volcanic eruptions. Grand Manan Island's flood basalt is also closely related to the North Mountain basalt of Nova Scotia. They certainly came from the same source, and may even be part of the same lava flows.

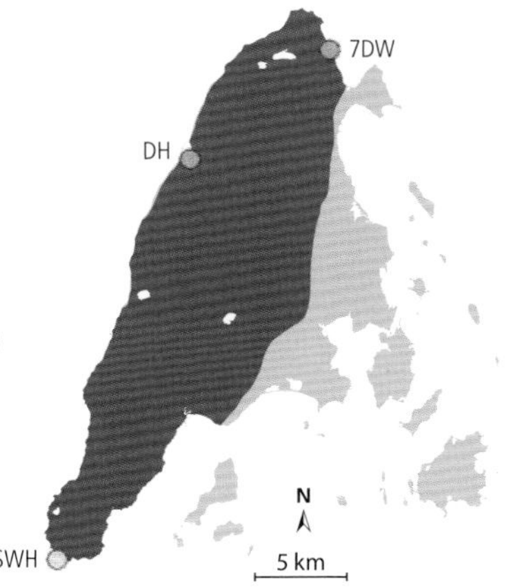

Also Nearby

Most of Grand Manan Island is made of basalt (purple in the map at right; Southwest Head, SWH). The site from which Grand Manan Island's basalt gets its official geological name is Dark Harbour (DH) on the west shore. A cliff face on the north shore and visible from the ferry is known as Seven Days Work (7DW) because several thick lava flows are visible there.

Exploring Further

Grand Manan Museum, 1141 Route 776, Grand Manan (N44.68690 W66.76467), www.grandmananmuseum.ca. The museum includes a permanent geology exhibit.

McHone, Greg. *Grand Manan Geology*. earth2geologists.net/grandmanangeology/ (extensive documentation of the local geology, including a field guide by the island's resident geologist).

The CAMP Site, https://www.auburn.edu/academic/science_math/res_area/geology/camp/CAMP.html (a collection of images and information about the igneous province).

Long Pond Beach is the site of several interesting outcrops in Anchorage Provincial Park on Grand Manan Island.

Faults, Folds, Flows

Ediacaran Rocks of Eastern Grand Manan Island

Geologists often describe Grand Manan Island as having a dual identity. The larger, western part is blanketed in flood basalt (site 8, Southwest Head). On the eastern side is a complicated array of much older rocks, the origins of which are still being debated.

In eastern Grand Manan Island and nearby islands, faults have brought numerous terrane fragments side by side. Some evidence suggests that most were once part of the mainland New River terrane. But even basic age information for some of the rocks has eluded researchers, and mysteries remain (see Also Nearby). If you read other books and papers about them, beware the phrase "assumed to be."

The Anchorage Provincial Park offers easy access to three significant features contributing to the island's complexity: (1) It provides the best available view of the major fault separating west from east. (2) The park's long shoreline exposes elaborately folded Ediacaran sedimentary rocks of mysterious provenance. (3) On its popular sandy beach, outcrops of volcanic rock display a variety of features formed by ash flows and other eruptions, also during the Ediacaran period.

Getting There

Driving Directions

Red Point site (1): From Route 776 near Seal Cove, turn east (N44.65578 W66.83816) onto Red Point Road and follow it to the shore. **Campground site (2)**: From Route 776 between Grand Harbour and Seal Cove, turn south (N44.67238 W66.80488) onto Anchorage Road and follow it into the park. Fork left (N44.66448 W66.80654) onto Long Pond Road and follow it to the shore. **Long Pond site (3)**: From Long Pond Road by the shore, turn sharply east (N44.65936 W66.80080) and follow the dirt road between the shore and the pond. About 800 metres from the campground, watch for the parking area on the left.

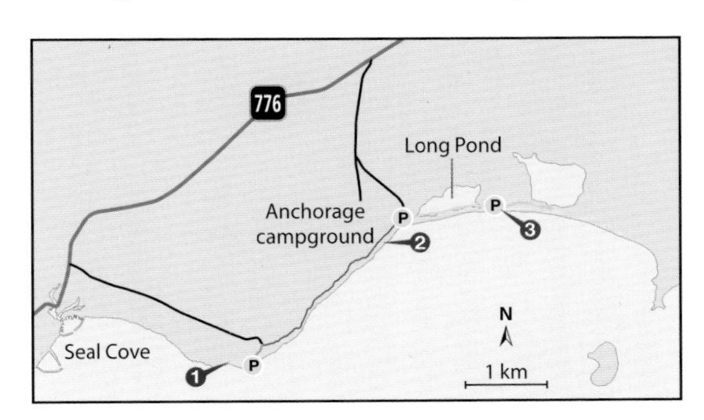

Where to Park

Parking Locations: Red Point (1): N44.64738 W66.81818
Campground (2): N44.65935 W66.80153
Long Pond (3): N44.66048 W66.79087

All these locations are gravel areas designated for parking. **(1)** is at the end of Red Point Road; **(2)** is on the west side of Long Point Road by the shore near the campground; and **(3)** is on the north side of the road just east of Long Pond.

Walking Directions

Red Point (1): From the southwest corner of the parking area, follow a short trail west and then south toward the shore. Walk down the slope onto the beach and continue west about 170 metres to the outcrop. **Campground (2)**: From the parking area walk straight onto the shore and then turn right (west) and continue about 280 metres to the outcrop. **Long Pond (3)**: The outcrop is on the beach in front of the parking area, but you may need to walk slightly east to access the beach, and then double back.

Notes

All the outcrops are on the shore and are best visited in low-tide conditions and quiet weather.

1:50,000 Map

Grand Manan Island 021B10

Provincial Scenic Route

Fundy Coastal Drive

On the Outcrop (1)

Groundwater seeping along the fault at Red Point has given rise to a line of small plants along the fault trace (see arrows) dividing basalt (left) from much older rocks (right).

Outcrop Location: N44.64773 W66.82065

At this site, note that the property above the fault is privately owned—please stay on the shore until you can re-enter the park.

No wonder the Red Point fault looks dishevelled. Geologists estimate that the basalt layers on the left slipped down about 2 kilometres along this slanting surface, relative to the ancient, reddish, metamorphosed volcanic rocks on the right. This type of fault, along which the rocks above slide down, is known as a normal fault.

Deformed basalt columns.

Immediately beside the fault the rocks on both sides are crumbly, broken up as they scraped against one another. About 30 metres west of the fault, its effects can still be seen. Once-vertical columns of basalt were deformed during fault movement and now look twisted or bent (detail). Farther west of the fault the columns are more nearly vertical.

Rock Unit

Dark Harbour Basalt;
Long Pond Bay Formation, Castalia Group

Bedrock Map

MP 2011-14

On the Outcrop (2)

On the shore below Red Point Trail, complex folds are easily visible thanks to contrasting colours of the rock layers.

Outcrop Location: N44.65742 W66.80353

This rock is a type of sandstone known as greywacke. Its minerals tell the story of an immature sediment deposited rapidly in large volumes. It contains rough, angular sand grains (no time to wear down), including bits of feldspar (which would have weathered to clay if not buried quickly), mixed with fine clay (sediment is normally sorted by size if it settles peacefully).

Conspicuous layering, caused by variations in grain size or mineral proportions, makes it easy to see the folds in the rock. Folds caused by tectonic activity are usually quite regular in form, aligned by forces pushing on them from a defined direction. These rather chaotic folds are not tectonic but sedimentary in origin, formed on an underwater slope when coherent but still pliable layers gave way and slumped downhill.

Given the rock's qualities, the setting for all these events was most likely a deep marine basin. With so many detailed features on view, the site brings such a setting to life. But where that basin was located, and what was being rapidly eroded to choke it with immature sediment, is not yet clear.

Rock Unit

Long Pond Bay Formation,
Castalia Group

Bedrock Map

MP 2011-14

On the Outcrop (3)

Volcanic rocks on Long Pond Beach display features including (**a**) spherulites, (**b**) rhyolite fragments, and (**c**) flow banding.

Outcrop Location: N44.66014 W66.79107

A fault hidden below the sand separates the rocks at this site from those at the previous outcrop. The rocks here all formed from volcanic ash resulting from explosive eruptions. Although the outcrops are limited and may change as storms rearrange the sand, it's a showcase of volcanic features.

One group of outcrops has a knobbly surface (photo **a**) with slightly flattened spheres about 2 to 4 centimetres wide. Known as spherulites, these bundles of microscopic crystals gather at a single point, looking a bit like a dandelion gone to seed. They formed after an eruption, as hot volcanic ash was cooling. In a different kind of lumpy outcrop (photo **b**), elongated, pinkish rhyolite fragments became embedded in volcanic ash during an eruption.

Some outcrops have smooth surfaces, but on closer inspection you'll notice subtle banding about 1 or 2 millimetres wide that forms complex swirling patterns (photo **c**). The contorted banding formed while stiff lava was still hot and slowly flowing or sagging. After an eruption, it might have taken about a week to cool and stop changing shape.

Rock Unit

Ingalls Head Formation, Grand Manan Group

Bedrock Map

MP 2011-14

FYI

- Zircon grains in volcanic rock form as the magma cools, so their age provides the age of the eruption. Results from the volcanic rocks at outcrop 3 indicate an age of about 618 million years.
- Zircon grains are found in sedimentary rock also, usually as tiny crystals much smaller than a grain of sand. They are unusually tough and resistant, so as a region is worn down by erosion, the resulting sediment may contain zircon grains from the whole area drained by its rivers. This gives the sediment a distinctive "signature" of zircon ages.

  A study of zircon ages from sedimentary rocks near outcrop 2 found that the youngest zircon grains they contain are about 600 million years old. That means the sedimentary rocks of outcrop 2 are likely to be somewhat younger than the volcanic rocks at outcrop 3.

Also Nearby

Mystery alert! Nearby White Head Island's eponymous headland began as pure quartz beach sand, now metamorphosed into gleaming white quartzite. The age signature of zircon grains preserved in the sand show that it was most likely derived from ancient rocks in what is now West Africa, an unusual finding. Most rocks in Ganderia probably originated in what is now South America.

To view this pure but puzzling quartzite, from the ferry terminal on White Head Island follow White Head Road around the harbour to its cul-de-sac termination (N44.62749 W66.72758).

White Head is named for this quartzite of uncertain origin.

Exploring Further

See site 8 (Southwest Head), for information about Grand Manan Museum and extensive online resources for the geology of Grand Manan Island.

The nature trail to Barnaby Head in New River Beach Provincial Park leads first to a small sandy cove.

Two Stories

A Deformed Granodiorite of the Brookville Terrane

Perhaps in your travels you have encountered an ancient, gnarled oak tree that seemed to tell a story. The granodiorite on Barnaby Head in New River Beach Provincial Park is similar. Originally formed late in the Ediacaran period, this intrusion survived a disaster of sorts during the Carboniferous period.

First picture the setting in which the rock at Barnaby Head originated. About 555 million years ago a subduction zone formed under the once-quiet margin of Ganderia. For about 30 million years, huge volumes of molten rock flooded upward from the subduction zone, rising through the crust below a chain of high volcanoes like those of today's Andes mountains.

The intrusion remained intact for 200 million years as Ganderia drifted toward and eventually collided with Laurentia. Then during the Carboniferous period, a slice of Ganderia's crust—including the rocks here and those at Dipper Harbour (site 12)—was thrust onto and then slid along an adjacent block of rocks. At Barnaby Head, the old granodiorite's figurative scars remain in the form of a rock fabric known as mylonite.

Getting There

Driving Directions

From Route 175 between Lepreau and Pocologan, about 1.5 kilometres east of the New River bridge, turn south (N45.13857 W66.52124) onto Haggertys Cove Road and drive past the entrance to New River Beach Provincial Park. Watch for Carrying Cove Road on the right (N45.13284 W66.52155) and follow it to the parking lot near the shore.

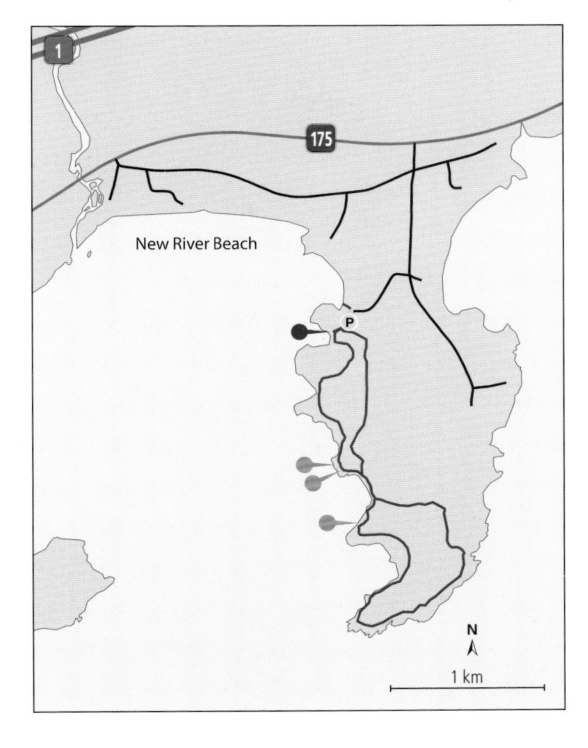

Where to Park

Parking Location: N45.13086 W66.52506

This is a large parking lot with access to the beach and trails.

Walking Directions

Exit the southwest corner of the parking lot and follow the short trail to a small sandy cove nearby. Facing the water, the cleanest outcrops are in the cliff on the right (north). As conditions allow, cross the beach to the outcrops.

For Barnaby Head trail, from the parking lot walk toward the shore, but follow the path left toward the woods instead of continuing onto the beach. Follow the trail about 1 kilometre to a boardwalk intersection. Bear right to a shoreline viewpoint and mylonite outcrops.

Notes

This site is located within New River Beach Provincial Park (www.tourism newbrunswick.ca). During the summer months, a fee is charged at the parking lot entrance.

1:50,000 Map

St. George 021G02

Provincial Scenic Route

Fundy Coastal Drive

On the Outcrop

A small headland at the north end of the beach provides extensive exposures of deformed granodiorite.

Outcrop Location: N45.13045 W66.52631

As you reach the cove and look right (north), the rocks at the end of the beach look like those of a typical felsic intrusion, blocky and hard. If you can find a relatively fresh surface, you'll see that the rock is light grey (the outcrop surface is mostly light brown due to weathering). That's because it is a granodiorite rather than a true granite and thus contains very little of the typically pink mineral potassium feldspar.

Looking at a clean, smooth surface you may see that, in places, the rock looks as if closely spaced parallel lines have been drawn across it. This feature is due to the realignment of dark minerals during deformation and is characteristic of mylonite.

Some quartz and feldspar grains resisted deformation and remain as rounded chunks up to 5 millimetres across, scattered among the finer grained, aligned dark minerals. As a result, in many places the rock has a mottled or banded appearance.

Mylonite fabric.

Rock Unit

Golden Grove Plutonic Suite

Bedrock Map

MP 2014-26

FYI

- Plate tectonic movements cause faults near the surface, where cold, brittle crust breaks suddenly. But at deeper, hotter levels, the Earth's crust is weaker. Instead of a fault, a shear zone forms—a wider zone of smeared-out rock. Typically, mylonite forms in a shear zone, as deformation breaks mineral grains into smaller and smaller pieces. The more intensely a mylonite is deformed, the more fine grained it becomes.
- The rock here and the granodiorite in Rockwood Park (site 15, outcrop 3) are both part of a group of 34 related igneous intrusions known as the Golden Grove Plutonic Suite (grey-green in the map at right). A pattern of similar ages and rock chemistry has allowed geologists to recognize this narrow band of intrusions more than 100 kilometres long—a prominent feature of the Brookville terrane.

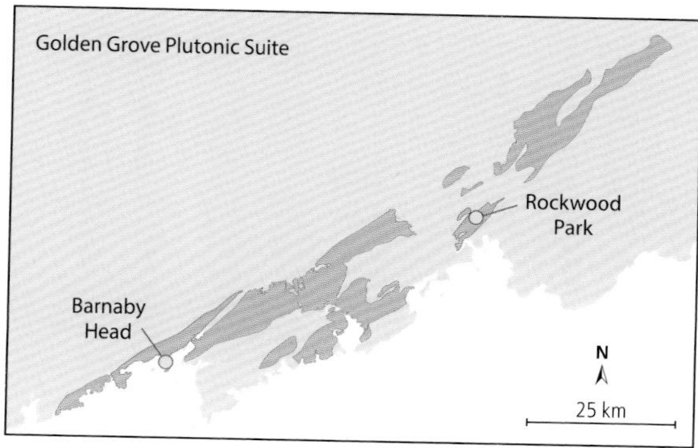

Also Nearby

The rocky shore accessible along a trail out to Barnaby Head has outcrops that are more deformed than those near the beach. About 1 kilometre from the beach (N45.12470 W66.52606 and N45.12451 W66.52569) are outcrops in which the minerals have been crushed during extreme deformation, forming a very fine-grained, banded mylonite. This happened during Carboniferous thrusting.

Sheared, mylonitic rock on Barnaby Head.

The Lepreau River tumbles over a sandstone ledge in scenic Lepreau Falls Provincial Park.

Yes, No, Maybe So

Sandstone That Is Probably Carboniferous

The Lepreau River flows across a variety of igneous rocks before reaching an isolated, fault-bounded sliver of red mudstone, sandstone, and conglomerate less than 1 kilometre wide. This sliver of sedimentary rock has given us a lovely spot in the form of Lepreau Falls; it has also given us a bit of geological intrigue.

The rocks' red colour, cross-bedding, and other features imply they were deposited by a river as part of an alluvial fan. In Atlantic Canada, similar red alluvial fan deposits formed during both the Carboniferous and the Triassic periods. For decades, the age of the rocks here remained unresolved with plenty of opinions on both sides, based on their resemblance to other New Brunswick rocks of known age.

In 1974 a sharp-eyed participant on a university field trip spotted fossil tracks in rocks of the Lepreau riverbed, but their age was ambiguous (see FYI). Over time, several lines of evidence—fossil plant spores, signs of tectonic stress, rock hardness, and a re-evaluation of the fossil tracks—converged on a consensus of Carboniferous age. It's a good example of how geologists figure things out by combining information from different disciplines.

Getting There

Driving Directions

From Route 1 between St. George and Saint John, take Exit 86 and follow Route 175 to Lepreau Falls Road (N45.17033 W66.46747), watching for signs to Lepreau Falls Provincial Park. From the intersection, travel east about 500 metres to the park entrance. Turn right (south) into the park's gravel road, drive past the washroom facility and park where the gravel widens in the picnic area.

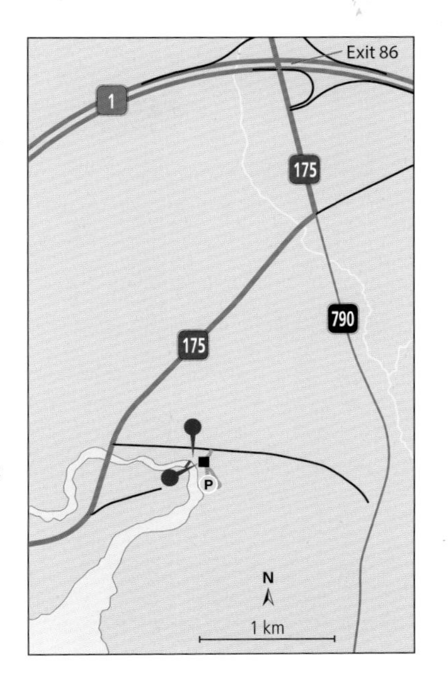

Where to Park

Parking Location: N45.16853 W66.46112

The park has no parking lot, but the gravel road widens for parking. Find a spot that allows traffic to pass through.

Walking Directions

From the park's gravel road, two boardwalks lead to lookouts with unobstructed views of the falls. As you face the washrooms, one boardwalk starts nearby on the left; it faces toward the falls. The other boardwalk starts farther along on the right, near Lepreau Falls Road. From it you can see straight across the falls and get a closer look at the rock layers.

Notes

Although water levels rarely affect access to the lookouts, rock layers in the falls are most visible when water levels are low.

Collection of fossils in New Brunswick requires a permit issued by the province.

1:50,000 Map

Musquash 021G01

Provincial Scenic Route

Fundy Coastal Drive

On the Outcrop

The face of Lepreau Falls exposes several thick layers of red sandstone.

Outcrop Location: N45.16988 W66.46201

In the rock face of the waterfall and immediately below the falls in the bed of the river, several layers of red sandstone are exposed. Above the falls, rocks in the rapids are finer-grained mudstone. The location of the falls here testifies to the rocks' hardness and resistance to weathering.

The layers are tilted slightly downward away from the falls, so from the lookout nearer the washrooms, you are looking at them roughly edge-on. From the lookout nearer the road, as you look across the river at the falls, the younger layers are on the right (upstream), and each successive downstream layer is older.

The sand and mud in these rocks were deposited in the channel of an ancient river, but several features (not all of them visible from the lookouts) indicate that the sediment layers were not continuously underwater. The fossil tracks, and even the red colour itself, are signs that the sand and mud were exposed to the air. The layers were most likely deposited during intermittent storms in a generally dry environment.

Rock Unit

Balls Lake Formation, Mabou Group

Bedrock Map

MP 2014-24

FYI

Badwater alluvial fan.

- Alluvial fans typically form in harsh, mountainous landscapes with little vegetation. The Badwater alluvial fan in Death Valley, California (photo at right), is a typical example.

- As the rock layers at Lepreau Falls formed, renewed uplift along Carboniferous faults sent rivers pouring onto the broad plains left by the final retreat of the Windsor Sea.

  When a river emerges from high ground onto a flat plain, it loses energy and must drop much of the sediment it carries. The river channel can become choked, causing the water to find another way downhill. Over time, numerous channels form, building up a wedge of sediment—an alluvial fan. Conglomerate from an alluvial fan of similar age can be seen at Hopewell Rocks (site 21).

- Experts agree that fossil tracks in the Lepreau riverbed were made by a four-legged creature about 11 centimetres long—but what kind? When? The initial study in 1977 proposed that they were the tracks of a Triassic reptile.

  Nearly 40 years later, in 2016, the tracks were re-examined. Armed with new information about early land dwellers, researchers found that the most likely candidate is a Carboniferous temnospondyl. Now extinct, temnospondyls were the ancestors of amphibians.

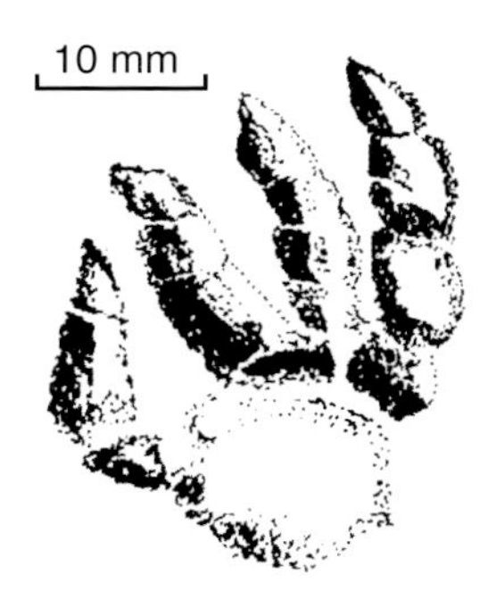

Sketch of temnospondyl track.

Exploring Further

Lepreau Falls is part of Stonehammer Geopark. For more information about the geopark's sites, programs, and amenities, visit https://www.stonehammergeopark.com.

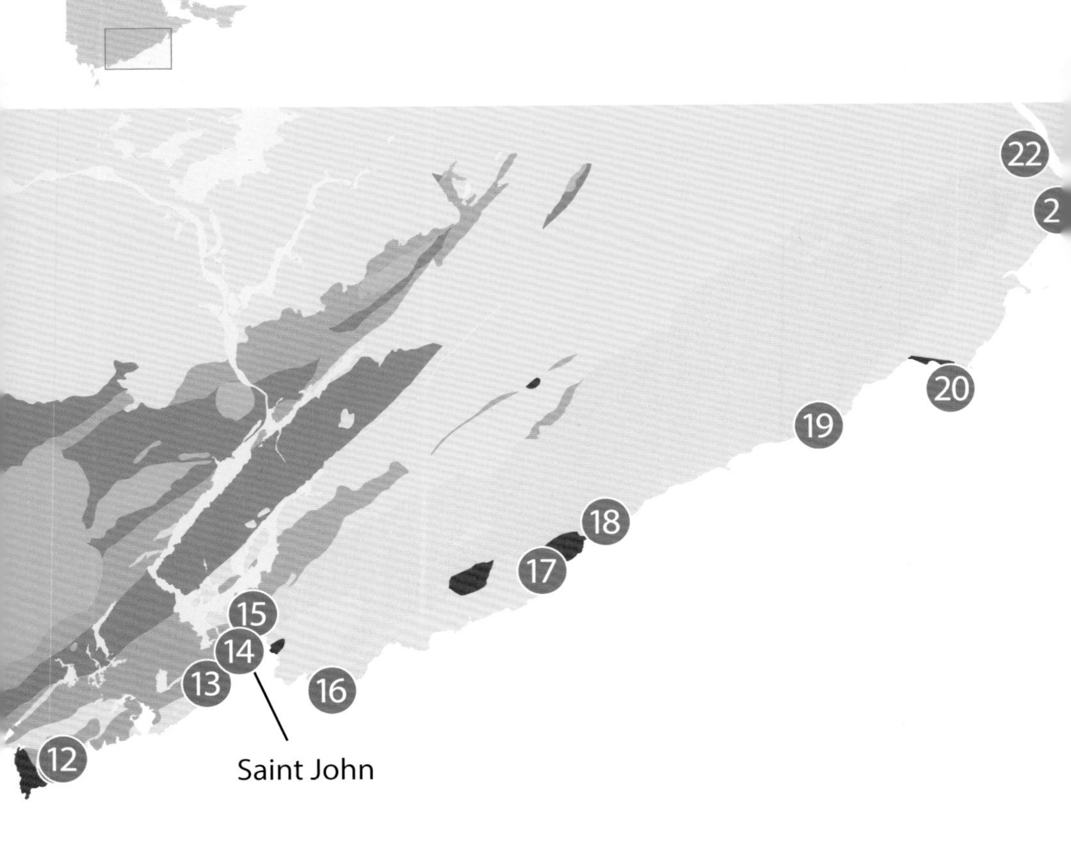

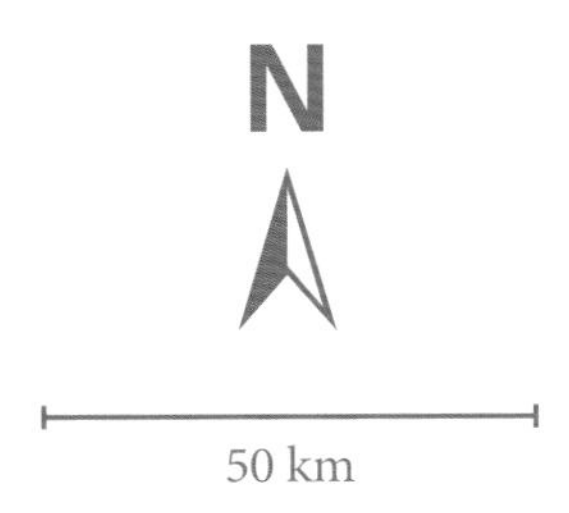

A. Foundations B. Collisions C. Pangaea D. Atlantic See pages 18–26

FUNDY – EAST

Tour 2 at a Glance

Tour 2 begins west of Saint John, then explores the city, including several sites in Stonehammer Geopark. Continuing east, the high sea cliffs and rolling hills of the province's southeastern highlands provide a scenic context for stories spanning nearly 500 million years of geologic time. The tour concludes along the broad Petitcodiac River, where the roots of Maritimes geoheritage run deep.

At these sites, you can explore ...

12	**Dipper Harbour**	Pyroclastic flows produced by a volcanic arc
13	**Taylors Island**	Messy side effects of a terrane collision
14	**King Square West**	Cambrian sandstone and shale of Avalonia
15	**Rockwood Park**	Ediacaran rocks of Ganderia and Avalonia
16	**Cape Spencer**	A fragment of Avalonia's foundation
17	**St. Martins**	Triassic sediments of the Fundy Basin
18	**Fundy Trail Parkway**	Evidence of Avalonia's dramatic past
19	**Point Wolfe**	Early Ediacaran volcanic rocks of Avalonia
20	**Cape Enrage**	River sandstone of the Coal Age
21	**Hopewell Rocks**	An alluvial fan deposit near a major fault
22	**Hillsborough**	Gypsum deposits of the Windsor Sea

Roadside scenery along Route 790 includes the rocky shoreline of Dipper Harbour's Back Cove.

Fiery Storm

Pyroclastic Flows Produced by a Volcanic Arc

Dipper Harbour is said to be named for its once-plentiful seasonal population of bufflehead ducks, a small diving species nicknamed "dippers." Naturalist J.J. Audubon travelled this way one spring in the 1830s and wrote that he had "found them exceedingly abundant on the waters of the Bay of Fundy."

Although named for a sign of ecological well-being—thriving flocks of little ducks—this cove is carved from rocks that originated in sudden, searing violence. About 550 million years ago, a volcanic arc formed above a subduction zone on the margin of Ganderia, where this site was located. The area may have resembled parts of the Pacific "ring of fire" of today, with many large volcanoes erupting dramatically.

Periodically, swirling masses of hot gas laden with volcanic ash, glowing globules of frothy lava, crystals, and other fragments rushed down the side of a volcano to settle here. Of course, that long ago only primitive life forms had evolved, and the land was barren. As you envision the ancient, catastrophic origin of these rocks, the dippers and their present-day cohabitants can stay safely tucked away in your imagination.

Getting There

Driving Directions

From Route 1 between St. George and Saint John, take Exit 86 or Exit 96 onto Route 790, which loops out around the shore between these exits. About 15 kilometres from either exit, just east of Dipper Harbour, watch for a gravel pull-off on the south side of the road, just as Route 790 passes near the shore in a broad, rocky cove (Dipper Harbour Back Cove).

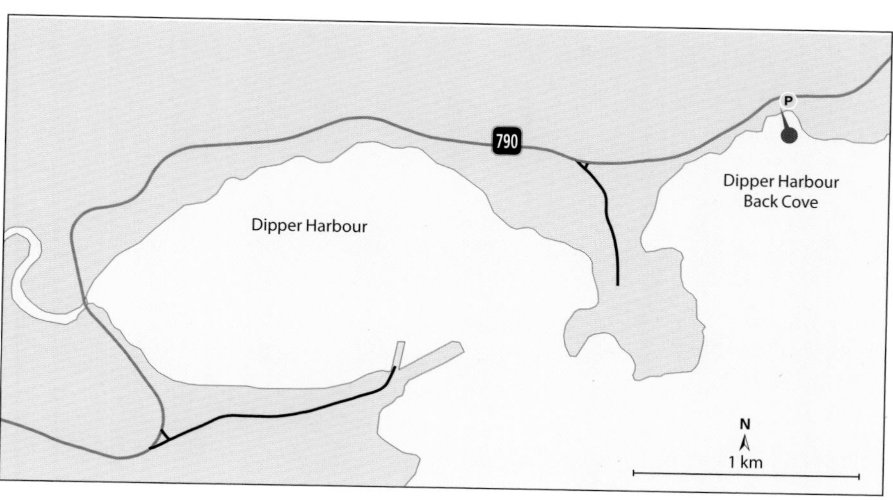

Where to Park

Parking Location: N45.10187 W66.40350

The parking location is a gravel pull-off on the south side of Route 790.

Walking Directions

From the pull-off, walk toward the water onto a rock pavement at the top of the outcrop. As you face the water, the left (east) side of the main outcrop forms a sloping ramp. As conditions allow, walk down to the beach. From there you can easily examine several areas of the outcrop.

Notes

This site is best visited at low tide, although parts of the outcrop are accessible even at high tide in otherwise normal conditions.

1:50,000 Map

Musquash 021G01

Provincial Scenic Route

Fundy Coastal Drive

On the Outcrop

The outcrop is wave washed during storms and offers many smooth, clean surfaces for viewing its volcanic features (see detail).

Outcrop Location: N45.10168 W66.40363

This pale grey rock is the volcanic equivalent of a granite intrusion, rich in quartz and feldspar, as you might guess from its colour. The texture of the rock signals its volcanic origins. Visible crystals and other eruption debris are embedded in a very fine-grained matrix of volcanic ash.

On the western side of the outcrop are some surfaces that display these features especially well. Visible crystals of quartz and feldspar, looking like very coarse salt, can be seen on the rock surface. In some places they have weathered more slowly than the fine matrix around them, so they stand out from the surface. Prior to the eruption these crystals had been floating in the liquid magma beneath the volcano.

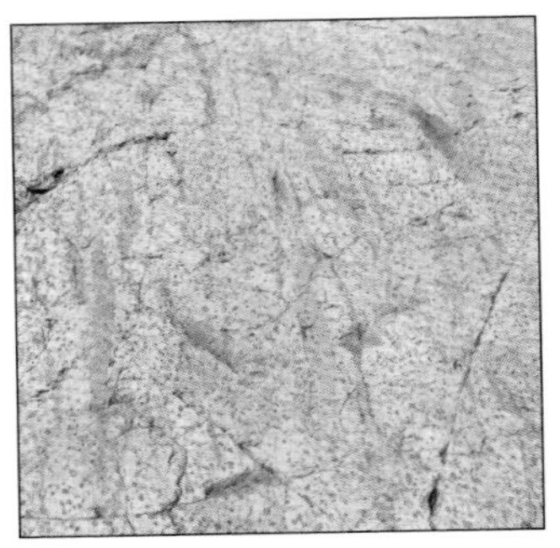

Pumice fragments in tuff.

In another shade of grey are irregular, narrow shapes up to 10 or more centimetres long. These are the flattened remains of pumice fragments—frothy globules of magma caught up in the roiling, hot ash cloud.

Rock Unit

Clear Lake Formation,
Dipper Harbour Group

Bedrock Map

MP 2001-30

FYI

- This site preserves a rare example of volcanic rocks associated in age and origin with the Brookville terrane's Golden Grove Plutonic Suite (see site 10, Barnaby Head, and site 15, Rockwood Park, outcrop 3). For all these rocks, the geologic setting was analogous to the modern-day Andes—with volcanoes erupting explosively at the surface and related intrusions forming below the surface.

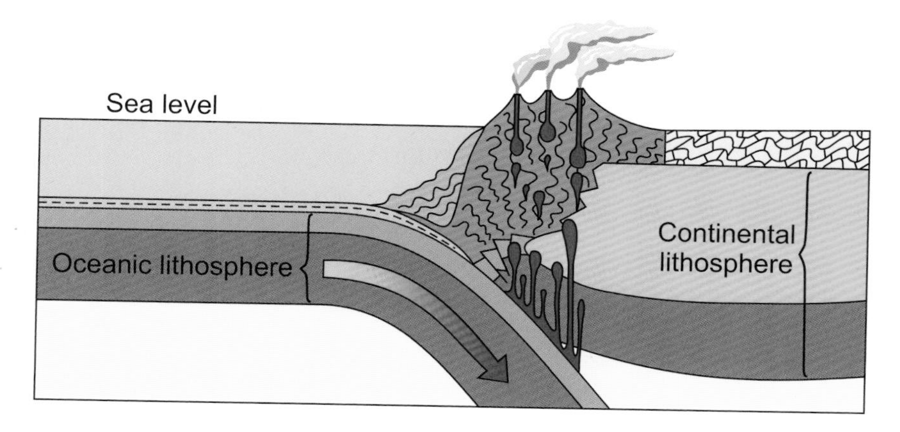

- Explosive eruptions are common in volcanoes above subduction zones because the magma typically contains water and is rich in silica. Silica-rich magma is viscous and does not flow easily, but water expands rapidly as it rises upward and is decompressed. It is this combination that leads to pyroclastic flows, as the thick magma is blown apart by expanding gas.
- The volcanic rocks around Dipper Harbour are allochthonous (not in their original location). During the Carboniferous period they were thrust northward onto Carboniferous sedimentary rocks by powerful fault movements. The Meguma terrane (now in parts of Nova Scotia) arrived in the area during that time and its motion probably triggered this disturbance. Signs of deformation related to this same cause can be seen at site 10, Barnaby Head; site 13, Taylors Island; and site 16, Cape Spencer.

Exploring Further

Audubon, J.J. "Buffle-Headed Duck." In *Birds of America*, vol. 6, pp. 369–373. New York: J.J. Audubon, 1843.

King, Hobart M. "Pumice." https://geology.com/rocks/pumice.shtml (detailed information about pumice in volcanic eruptions).

As seen from Manawagonish Road west of Saint John, Taylors Island lies between marshland and the Bay of Fundy.

Hot and Squishy

Messy Side Effects of a Terrane Collision

Taylors Island in the Bay of Fundy is connected to the mainland southwest of Saint John by Saints Rest Beach, and the island's northern end is surrounded by a marsh along Manawagonish Creek. The island, beach, and wetland are all part of Irving Nature Park, which extends farther east to include Sheldon Point. Founded primarily to protect the area's ecosystems, the park features a rocky shoreline that also provides geology enthusiasts with plenty of outcrops to explore.

The rocks on Taylors Island formed in a complex tectonic setting. A terrane known as Meguma (now seen only in Nova Scotia) had collided with Avalonia, and movement between them continued as they pressed and scraped past one another along a set of major strike-slip faults.

Here, silt and sand accumulated in a small basin along the fault system. While the sediment was still wet and soft, basalt lava erupted into the layers, creating what can only be described as a bit of a mess. To geologists it is known as peperite. Taylors Island provides some excellent examples of all three rock types: the sand- and siltstone, basalt, and peperite.

Getting There

Driving Directions

From Route 1 in Saint John, take Exit 119 and travel south on Bleury Street about 500 metres to Sand Cove Road. You may see signs for Irving Nature Park or for the Stonehammer Geopark. Turn (N45.24661 W66.09323) right (southwest) and follow Sand Cove Road about 4 kilometres to **Saints Rest Beach** (**1**). From the beach, follow the unpaved park road around Taylors Island to a small **cove** (**2**); the park road is a one-way loop of 6.4 kilometres.

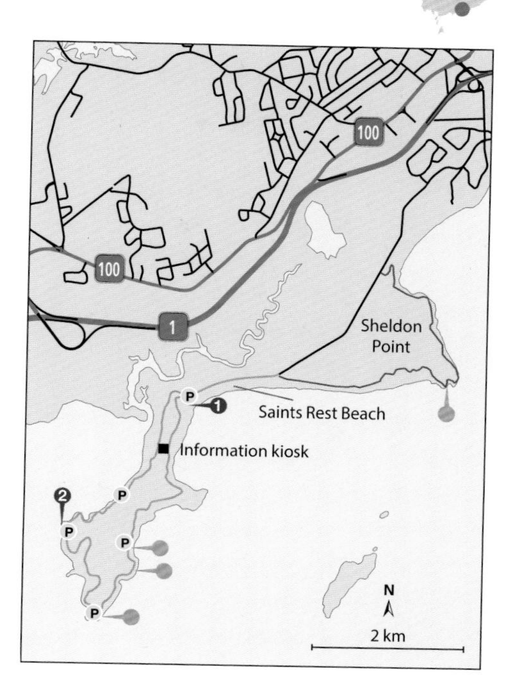

Where to Park

Parking Locations: Saints Rest (1): N45.22395 W66.12759
Cove (2): N45.21232 W66.14208

The nature park provides several parking areas on Taylors Island, ranging from large gravel lots with washrooms to small gravel pull-offs by the road.

Walking Directions

Saints Rest (**1**): From the parking area, in low-tide conditions walk onto the beach. Look right as you face the water to view rock surfaces. **Cove** (**2**): From the parking area walk onto the shingle beach to view outcrops both right and left along the shore.

Notes

A large map of Taylors Island is on display beside the park road at Saints Rest Beach. Vehicle access is seasonal. The island, as part of Irving Nature Park, is a protected natural area. Removal of any material, even beach stones, is prohibited.

1:50,000 Map

Musquash 021G01

Provincial Scenic Route

Fundy Coastal Drive

On the Outcrop (1)

Contacts between grey basalt and reddish siltstone swirl and blend in an outcrop of peperite on Saints Rest Beach, Taylors Island.

Outcrop Location: N45.22309 W66.12772

Two very different rock types mingle in this outcrop to form peperite. The left side is dominated by greenish grey basalt. On the right side are steeply tilted layers of dark and pale pink siltstone. In some places where they meet (see detail), the basalt forms bulbous shapes against the siltstone. Elsewhere, the boundary seems gradual, with one rock type fading into the other; or you may see irregular fragments of one rock type scattered through the other. These are signs that the sediment was still soft when the lava erupted.

The outcrop also preserves some classic signs of glaciation. This happened quite recently in geologic terms—about 10,000 years ago. The smooth, round curves of the rock surface and the fine horizontal grooves that cross it are both the result of a massive ice sheet that moved over the landscape. Rock dust in the ice sculpted and polished the outcrop surface, while pebbles in the ice scraped along making the grooves.

Peperite.

Rock Unit

Taylors Island Formation, Lorneville Group

Bedrock Map

MP 2001-33 (rev. 2014)

FYI

- The term "peperite" has a rather cosmopolitan derivation. It comes from an earlier form, "peperino," borrowed in 1827 from the Italian by an English geologist studying volcanic rocks in central France. The rocks there reminded him of coarsely ground pepper.

  For more than a century, the term was applied to various kinds of volcanic rock based on some resemblance to pepper. It was only a few decades ago that geologists agreed peperite should refer only to rocks formed in a particular way (that is, lava mixing with wet sediment) rather than to a peppery appearance.

- Basalt (or gabbro) and sandstone or siltstone are found together in other parts of New Brunswick (see sites 2, St. Andrews; 3, Ministers Island; and 25, Currie Mountain), although not as peperite. This combination of rock types formed during the long history of Devonian and Carboniferous strike-slip faulting in the region.

Also Nearby

For more views of the red mudstone, park on the east side of the island (N45.21100 W66.13465). From the southeast corner of the parking area, cross the road to a pavilion and observation deck. From the deck, you can see mudstone outcrops on the beach below.

From the south side of the same pavilion, follow the Heron Trail south about 300 metres for additional views of mudstone and peperite (for example, N45.20855 W66.13370). The rocks are visible from the trail. Please help preserve the island landscape—stay on established footpaths.

Peperite visible from Heron Trail.

On the Outcrop (2)

A low outcrop of basalt stands at the north end of this pebble beach on the west side of Taylors Island.

Outcrop Location: N45.21250 W66.14255

Much of the rock that frames this pebble beach has a greenish cast—what some geologists would informally call greenstone. When basalt is altered by interaction with circulating fluids, its very dark green, nearly black, pyroxene crystals break down and form minerals like chlorite and epidote that are lighter or brighter green.

Many of the pebbles on the beach are also greenstone, uniformly greenish grey. Others are gabbro, with larger, visible light and dark crystals. Some crystals are bright green epidote.

Mineral-rich fluids passing through the basalt while it was being deformed caused numerous veins to form in the rock. In some outcrops you may notice an unusual type of mineral growth in veins. The veins look like tiny ladders because the mineral grains stretch from one side to the other (see detail). This happens when the vein fissure opens slowly, giving the crystals time to grow longer.

Veins in basalt.

Rock Unit

Taylors Island Formation, Lorneville Group

Bedrock Map

MP 2001-33 (rev. 2014)

FYI

The Partridge Island Block

- Geologists consider the rocks exposed on Taylors Island to be part of the Partridge Island block, which formed during an unusual tectonic upheaval early in the Carboniferous period.

  The rocks of Taylors Island are only slightly metamorphosed. But close beside them (see map) are small areas of granitic and metamorphic rock (mylonites) of similar age, but much more deformed. These rocks formed about 15 kilometres deep in the crust, then were rapidly transported upward in a fault zone. This violent juxtaposition of deep and shallow rocks is known as a metamorphic core complex, more commonly seen in large mountain ranges like the Rockies. The cause of this chaos appears to have been the oblique collision of Meguma with Avalonia.

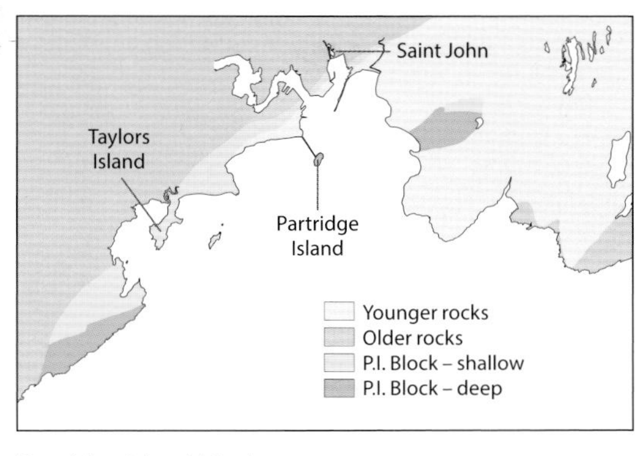

Partridge Island block.

- Some of the mylonite is well exposed along Bayside Drive at and north of the intersection with Hewitt Road (N45.24548 W65.98685; this is along the route to site 16, Cape Spencer).

Also Nearby

All the rock outcrops around the southern tip of Taylors Island are basalt that has been altered by circulating, mineral-rich fluids. To see another exposure with abundant mineral veins, park in the large lot at the southern tip of the island (N45.20508 W66.13806). Exit from the southwest corner, then walk south across the park road. From the lookout platform, you'll see red siltstone and dark, greenish basalt on the beach. A trail through the woods leads east to a barren point of altered basalt (N45.20448 W66.13761). In the outcrops are numerous veins of white quartz, white calcite, and bright green epidote (photo at right).

Quartz and epidote veins.

The rocks on nearby Sheldon Point are similar, related rocks. From the trail head along Sand Cove Road at the east end of Saints Rest Beach (N45.22572 W66.11720), follow the footpath east about 1.7 kilometres to the point.

Capped by pinkish grey, fine-grained sandstone and dark grey shale, the hilltop in King Square West in Saint John offers a broad view of the city.

Wave Washed

Cambrian Sandstone and Shale of Avalonia

King Square West is a quiet community park conveniently near the Bay Ferries terminal. What's more, it is host to sedimentary rocks that formed during the Cambrian period, an important interval in the history of Avalonia.

The bridge at Reversing Falls, less than 1 kilometre away, is visible from the high ground in the park and marks the boundary between rocks of Avalonia and those of Ganderia. But when the sandstone and shale in King Square West were being formed on the shores of Avalonia 500 million years ago, Ganderia was far away, across an ocean hundreds or even thousands of kilometres wide.

Volcanic activity that defined Avalonia for millions of years had largely ceased. A rise in sea level (or reduced elevation of the land) caused much of the region to be covered by a shallow sea. Features in the sandstone layers in the park indicate that the shore here was breezy, perhaps stormy, leaving many signs of wave action. The shale layers suggest an ecosystem busy with primitive animal life burrowing in the mud—appropriate, since these rocks formed during the Cambrian explosion, a time of rapid evolutionary change.

Getting There

Driving Directions

From Route 1, take Exit 120 (as if to the Digby ferry).

The eastbound exit ramp empties onto Ludlow Street. Follow Ludlow two blocks to Duke Street and turn right (southwest). Follow Duke Street one block to the park.

The westbound exit ramp empties onto Market Place. Follow Market Place two blocks to Duke Street and turn right (southwest). Follow Duke Street two blocks to the park.

Where to Park

Parking Location: N45.25814 W66.07675

The widened lane beside the park on Duke Street offers convenient parking if available.

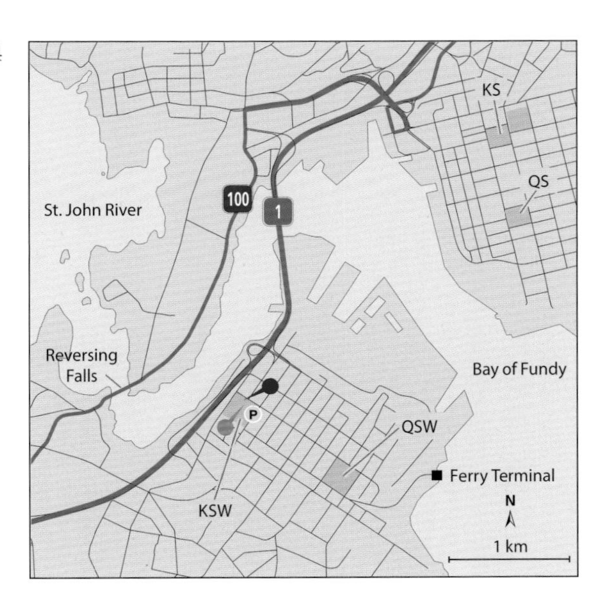

Walking Directions

From the parking location, follow the sidewalk downhill to Watson Street, then along Watson to King Street. (Watch for oncoming traffic or cross to the sidewalk on the far side of Watson.) The main outcrop location is across from the end of King Street. For another view of this outcrop, follow the footpath in the park from Duke Street to outcrops on the high ground.

Notes

Saint John boasts two King Squares (two Queen Squares as well), one in the old downtown (northeast of the river) and one in Saint John West (southwest of the river). If using a GPS navigation system, be sure to use the whole name, King Square West, and verify that the map is indicating the correct destination.

1:50,000 Map

Saint John 021G08

Provincial Scenic Route

Fundy Coastal Drive

On the Outcrop

Beside a grassy niche along Watson Street, steeply tilted layers of fine-grained sandstone, siltstone, and shale are well exposed.

Outcrop Location: N45.25889 W66.07733

For additional outcrops, follow the park footpath from Duke Street onto the high ground (N45.25857 W66.07740).

These sedimentary layers began as deposits of fine sand, silt, and mud (now sandstone, siltstone, and shale). The sandy layers can be recognized as lighter in colour, blocky, and more prominent because they resist weathering better than the other rock types.

The edges of the sandstone layers are generally too weathered to reveal details, but ripple marks from wave action are common here. Some shale layers have a lumpy appearance. The mud from which they formed was inhabited by worms and other primitive animal life that churned through the sediment in search of food, a process known as bioturbation.

Later deformation has tilted the layers, exposing a sequence of ages—older rocks at the right, younger at the left as you face the outcrop along Watson Street (see photo). Bright white veins of quartz cutting across some parts of the outcrop are also the result of deformation. Such veins tend to be more abundant in areas like this one, with plenty of sandstone layers as a source for the silica from which quartz forms.

Rock Unit

King Square West Formation, Saint John Group

Bedrock Map

MP 2007-5 (rev. 2014)

FYI

- King Square West is the type locality for the King Square Formation, the thickest and best exposed rock formation in the Saint John Group. The identification and naming of rock formations is a formal process within the geological community. A type locality is chosen as the "official" representative of a rock formation so that it can be correctly recognized elsewhere.
- The location of Avalonia during the Cambrian period is still being investigated. Signs of Ediacaran rifting suggest separation from Gondwana, but rather than separating, Avalonia may have slid along the edge of Gondwana instead.

  Evidence for this idea comes from the ages of zircon grains in sedimentary rocks of Avalonia (including the King Square Formation). The pattern of ages suggests that the source of the sediment changed during the Cambrian period. Zircon ages in older sedimentary layers match the pattern of ages from the Amazonian region of Gondwana. But zircon ages in younger sedimentary layers better match the pattern from the West African region of Gondwana.

Also Nearby

King of the castle. From the high ground in King Square West you can see several geological landmarks associated with Stonehammer Geopark. Looking west, just left of the Irving paper mill is Reversing Falls (N45.25800 W66.08830, see photo). Looking north, past the highway bridge lie Fort Howe (N45.27642 W66.07300) and the hills of Rockwood Park (site 15).

Exploring Further

For more information about Reversing Falls, Fort Howe, and other sites, programs, and amenities affiliated with Stonehammer Geopark, visit https://www.stonehammer geopark.com.

Rockwood Park in Saint John was designed by a famous landscape architect, Calvert Vaux, who favoured rustic views like this rocky woodland on Fisher Lakes.

Two Worlds

Ediacaran Rocks of Ganderia and Avalonia

At 870 hectares, Saint John's Rockwood Park is one of Canada's largest urban parks. Founded in 1894, it includes 50 kilometres of wooded trails, lakes, a campground, and other attractions. The park's rock outcrops preserve some of New Brunswick's oldest rocks and also a terrane boundary separating parts of Ganderia and Avalonia.

About 600 million years ago, this part of Ganderia was a stable, quiet continental margin. But about 550 million years ago a dramatic shift of tectonic plates caused a subduction zone to form beneath it. The sedimentary rocks of the margin were deformed and metamorphosed, as huge volumes of molten rock flooded upward, intruding them. This chapter in Ganderia's story is preserved here as part of the Brookville terrane.

In the park are ancient sedimentary rocks of Ganderia's stable margin (outcrop 1), a related sediment-derived gneiss (outcrop 2), and a 540-million-year-old granodiorite that intruded them (outcrop 3). A significant fault also cuts through the park. South of it, along Lake Drive near the Interpretive Centre, is another world: rocks of Avalonia, which arrived beside Ganderia after a long, complex journey.

Getting There

Driving Directions

If travelling west on Route 1, take Exit 125 and keep left on the exit ramp for a left turn onto Seeley Street. Watch for blue-and-white camping signs. Use the right lane on Seeley and take the first right onto Mount Pleasant Avenue. The entrance to the park (N45.28772 W66.05832) is on the right.

If travelling east, from Route 100 north of downtown watch for blue-and-white camping signs and exit (N45.28326 W66.05260) onto Crown Street, which becomes Mount Pleasant Avenue. Follow Mount Pleasant to the park entrance and turn right.

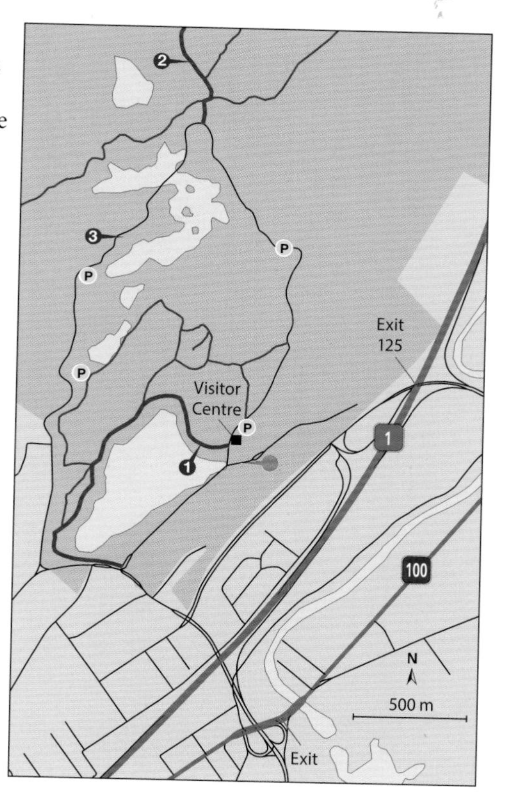

Where to Park

Parking Locations:

Site 1: N45.29185 W66.05365
Site 2: N45.29713 W66.05225
Site 3: N45.29626 W66.05993

All the parking locations are sizeable gravel lots adjacent to Fisher Lakes Drive. Smaller pull-offs may be accessible at times.

Walking Directions

Site 1: From the parking lot walk south past the Interpretation Centre, then cross Fisher Lakes Drive. Beside Lily Lake, follow the Cedar Trail to the outcrop locations (400 metres round trip). **Site 2**: Follow Fisher Lakes Drive north and west about 500 metres to the start of the Zoo Trail (N45.30061 W66.05558). The trail jogs right, then left, and crosses under a power transmission line before reaching the outcrop location about 250 metres from the trail head (1.5 kilometres round trip). **Site 3**: From the parking location, walk north on Fisher Lakes Drive about 250 metres to outcrops in front of the provincial displays for Alberta and British Columbia (500 metres round trip).

1:50,000 Map

Saint John 021G08

Provincial Scenic Route

Fundy Coastal Drive

On the Outcrop (1)

Limestone and sandstone of Ganderia's ancient continental shelf have been metamorphosed to form (**a**) marble and (**b**) quartzite.

Outcrop Location: N45.29129 W66.05492

The waypoint given is for the marble outcrop along Cedar Trail. About 75 metres farther along the trail are outcrops of quartzite.

The marble outcrops generally have a rounded surface because the carbonate minerals are slightly soluble in water, and so are subject to chemical weathering in the maritime climate. You may find that they have a slightly bluish cast. Look for intricately folded traces of contrasting light and dark grey on the outcrop surface. When carbonate rocks are subjected to stress, they become highly plastic and are easily distorted. The complex shapes are the remains of limestone bedding (deposited about 600 million years ago), contorted by deformation and low-grade metamorphism in the subduction zone (550 to 525 million years ago).

The quartzite occurs as discontinuous pods within the marble. During subduction-related deformation the brittle quartzite beds had no support from the easily deformed marble. They were broken up and engulfed by the pastelike flow of the marble around them. The original sandstone bedding is still preserved within the pods, although tilted by deformation. The quartzite outcrops have a blocky appearance, with sharp edges where beds have fractured and broken.

Rock Unit	**Bedrock Map**
Ashburn Formation, Green Head Group	MP 2007-5 (rev. 2014)

On the Outcrop (2)

In an outcrop of Brookville gneiss along the Zoo Trail in Rockwood Park, the foliation is masked by weathering but recognizable (see red line).

Outcrop Location: N45.30247 W66.05627

Despite the outcrop's weathered appearance, on close inspection you can still see a variety of features. You may first notice lighter coloured, pinkish granitic dykes cutting across the outcrop. They are younger and not part of the gneiss itself.

From the trail you see the gneiss layers end-on. They are steeply tilted, as seen in the photo above (a red line is drawn parallel to the layering). A few surfaces on the outcrop are clean, displaying the narrow bands of contrasting colour. The rock surface is ridged in alignment with the layering, since various layers weather differently.

Geologists have found aluminum-rich metamorphic minerals like cordierite and sillimanite in the gneiss. These indicate that the gneiss formed from sediments containing clay minerals (the source of the aluminum). They also tell how deeply the sediment was buried to form the gneiss (about 8 kilometres) and how hot they became (up to 650°C). Most geologists currently think the marble and quartzite (outcrop 1) and this gneiss were all formed from a continuous, related sequence of sedimentary rocks along Ganderia's margin.

Rock Unit	Bedrock Map
Brookville Gneiss	MP 2007-5 (rev. 2014)

On the Outcrop (3)

One of several outcrops of Rockwood Park granodiorite along Fisher Lakes Drive is located near the Canada 150 monuments for Alberta and British Columbia.

Outcrop Location: N45.29735 W66.05874

The Rockwood Park granodiorite is part of the Golden Grove Plutonic Suite (see site 10, Barnaby Head, FYI), which arose from a subduction zone under Ganderia's margin early in the Cambrian period, about 540 million years ago.

Given such a dramatic context, the granodiorite shows signs of a relatively placid history. It is quite coarse grained, meaning it cooled at a moderate pace. Its minerals are not strongly aligned, either, meaning it has never been strongly deformed.

Granodiorite.

Your best chance to see these details on a clean surface will come from looking around the lower portions of the outcrop. In recessed areas that are sheltered from the weather, you'll be able to see the obviously speckled appearance consisting of white feldspar, light grey or white quartz, and black hornblende.

Rock Unit

Rockwood Park Granodiorite

Bedrock Map

MP 2007-5 (rev. 2014)

FYI

- Running nearly parallel to Lake Drive South and the shore of Lily Lake is a significant fault. It separates rocks of the Brookville terrane (outcrops 1 to 3; part of Ganderia) from volcanic rocks of the Coldbrook Group (part of Avalonia; see site 18, Fundy Trail Parkway).

  The rocks of Avalonia can be seen between the intersection of Lake Drive and Fisher Lakes Drive and the entrance to the campground (N45.29080 W66.05384). The high, greenish grey outcrop is a fine-grained volcanic rock known as dacite.

Dacite.

- Avalonia and Ganderia were both still part of Gondwana when the rocks in Rockwood Park formed between 600 and 540 million years ago. They were widely separated by the Rheic Ocean during the Cambrian and Ordovician periods. During the Silurian period, Avalonia first came into contact with Ganderia, but the juxtaposition you see in the park today may not have been complete until sometime in the Carboniferous period, when strike-slip movement among New Brunswick's ancient terranes finally ceased. One long journey!

Also Nearby

Along the northern section of Fisher Lakes Drive in Rockwood Park is a series of 14 polished stone monuments, one for Canada and one for each province and territory. Erected in celebration of Canada 150, each monument is made of rock emblematic of its region. The location of the New Brunswick monument, made of St. George granite (see site 4), is marked with an orange dot on the map at right (N45.29899 W66.05489). Look for these monuments in gravel pull-offs and alcoves along Fisher Lakes Drive.

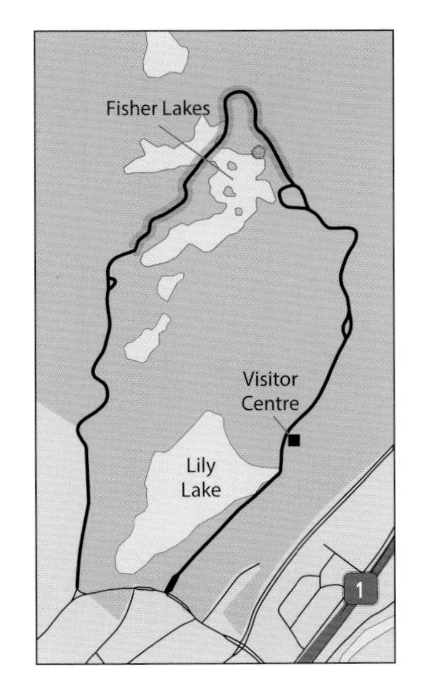

Exploring Further

For more information about Rockwood Park trails, attractions, and programs, visit http://www.rockwoodpark.ca and https://www.stonehammergeopark.com.

On a clear day, the expansive view from Cape Spencer includes ships bound to and from Saint John, as seen in the background, at right.

A

Look West

A Fragment of Avalonia's Foundation

How fitting that Cape Spencer is literally at the end of the road. As part of the granitic foundations of Avalonia, Cape Spencer marks the most westerly outcrop of its kind in all of Atlantic Canada (see FYI). More than 500 kilometres to the southwest—down the Bay of Fundy, across the Gulf of Maine—is its nearest kin.

The Dedham granite near Boston, Massachusetts, forms an equivalent part of Avalonia's foundation in New England. During the last ice age, glaciers transported fragments of Dedham granite to the shore along Cape Cod Bay. And so it happened that Plymouth Rock, landing place of the Pilgrims, is derived from a granite similar in composition and origin to the rock at Cape Spencer.

Early in the Ediacaran period a subduction zone formed along the margin of Gondwana, where Avalonia originated. Magma rose toward the surface, forming granitic intrusions and explosive volcanic eruptions. Rocks formed in this setting have been found all across Avalonia, from New England to the Maritimes and Newfoundland. They provide one way of recognizing the microcontinent's scattered remains.

Getting There

Driving Directions

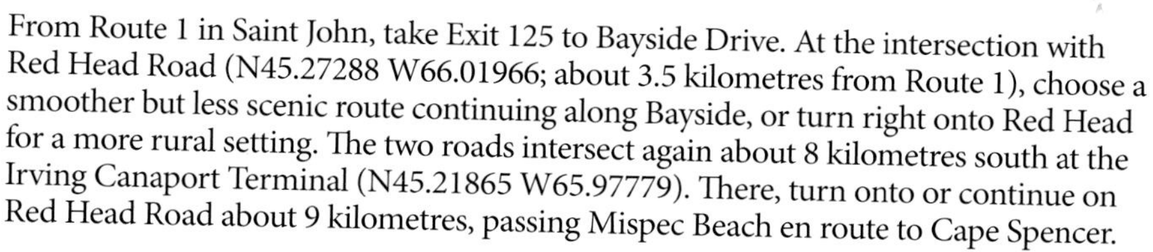

From Route 1 in Saint John, take Exit 125 to Bayside Drive. At the intersection with Red Head Road (N45.27288 W66.01966; about 3.5 kilometres from Route 1), choose a smoother but less scenic route continuing along Bayside, or turn right onto Red Head for a more rural setting. The two roads intersect again about 8 kilometres south at the Irving Canaport Terminal (N45.21865 W65.97779). There, turn onto or continue on Red Head Road about 9 kilometres, passing Mispec Beach en route to Cape Spencer.

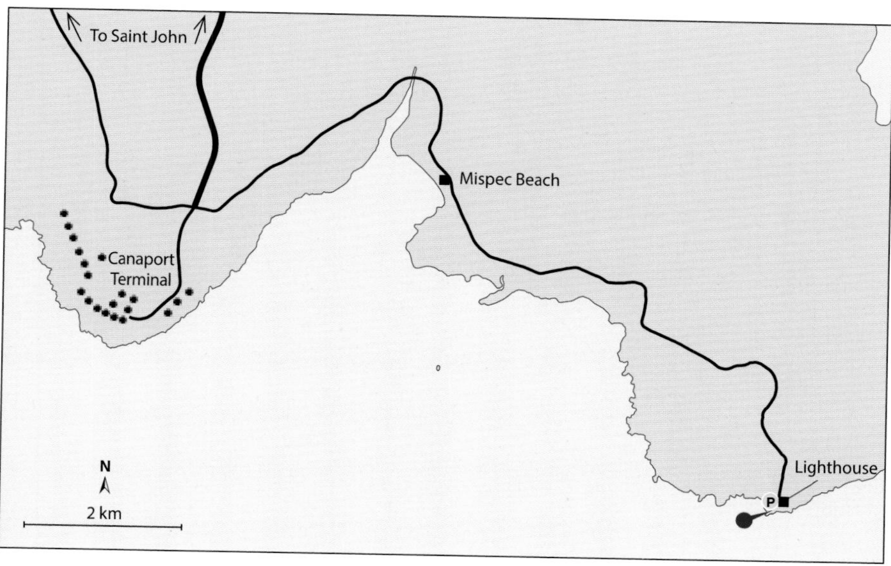

Where to Park

Parking Location: N45.19528 W65.91019

This is a small gravel pull-off west of Red Head Road (on the right as you arrive).

Walking Directions

From the parking location, with the lighthouse on your left, walk to the end of the road and follow a footpath about 20 metres to a nearby outcrop on the right.

Caution

In foggy weather, a loud horn may sound at this location.

1:50,000 Map

Cape Spencer 021H04

Provincial Scenic Route

Fundy Coastal Drive

On the Outcrop

Dark veins of chlorite and white veins of quartz cut through the deformed granite.

Outcrop Location: N45.19503 W65.91031

Geologic maps aptly describe this granite as "pervasively altered." Long after it formed, the granite was caught up in a fault system that affected many other rocks along New Brunswick's Fundy coast (see site 12, Dipper Harbour, FYI). Due to this deformation, the granite now has a noticeable mylonitic fabric.

In an unaltered granite you would expect to see flakes of biotite or blocky crystals of amphibole. But none of these have survived. They have all been replaced by green-black chlorite, veins of which appear as dark gashes or smears in the rock. You may also see white veins and pods of quartz.

Deformed granite.

The brownish, stained appearance of the outcrop is due to fluids carrying iron, manganese, and other metals, which flowed through the rock as it was deformed. On some surfaces, dark orange-brown specks resemble a mineral "rash." These mark the location of magnetite or pyrite grains, which rust when weathered.

Rock Unit

Millican Lake Granite

Bedrock Map

MP 2004-108

FYI

- The granite at Cape Spencer is related to volcanic rocks of similar age in Fundy National Park at Point Wolfe (site 19). Details of their chemical composition show that the rocks from both sites formed from magma generated above a subduction zone. All regions of Avalonia (light grey-green in the map below) include rocks in this age range.

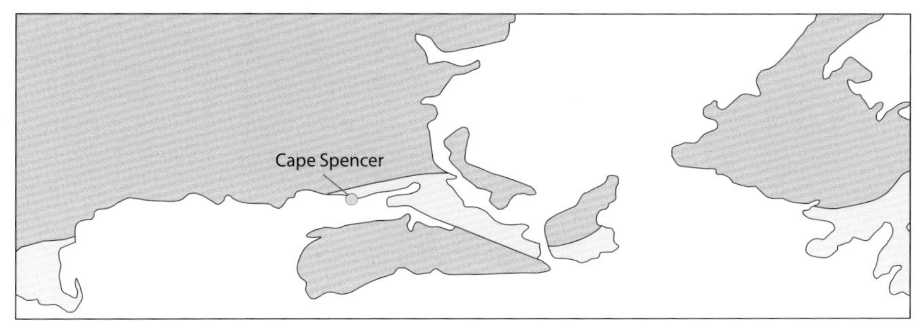

Avalonia in New England and Atlantic Canada.

- As described in the New Brunswick Bedrock Lexicon, the outcrop here is part of a larger intrusion that was "tectonically dismembered into at least nine individual thrust slices" during fault movements late in the Carboniferous period. Each slice is surrounded by a zone of intense deformation (see Also Nearby).
- During one or more phases of deformation, fluids flowing through the rocks near Cape Spencer left modest gold deposits in the area. First noted along the coastal cliffs of the cape during the 1950s, the deposits were more fully explored during the 1980s. An open-pit mine operated for a few years, less than 1 kilometre from the lighthouse.

Also Nearby

Near the parking location at the end of Red Head Road, beside the gravel lane leading around the back of the lighthouse, the rocks do not look like Cape Spencer granite. They are mylonite, sheared rocks in which the mineral grains have been crushed and smeared out during intense deformation. It's difficult to tell what the original rock was. It may have been granite or a type of feldspar-rich sandstone called arkose—both occur nearby.

Mylonite.

Exploring Further

New Brunswick Department of Energy and Resource Development. "Cape Spencer Mine." *Mineral Occurrence Database.* https://www1.gnb.ca/0078/GeoscienceDatabase (follow the database link and search for deposit name "Cape Spencer").

Grey conglomerate and brick-red sandstone form cliffs at the St. Martins sea caves east of Saint John.

Pangaea's Heartland

Triassic Sediments of the Fundy Basin

This scenic and geologically interesting site just outside St. Martins provides unique public access to New Brunswick's scarce outcrops of Triassic rocks in the form of cliffs and sea caves. Part of Stonehammer Geopark, the site also serves as a gateway to the rugged coastal region to the east, recently made accessible by the construction of the Fundy Trail Parkway (site 18).

The caves are carved in tilted red sandstone layers. Lying above the sandstone along an easily visible unconformity is a thick sequence of grey conglomerate layers. The conglomerate mainly consists of innumerable round cobbles and boulders, most of them quartzite from an unknown source, in a scarce matrix of fine sand.

In contrast to the relatively lush and temperate conditions along the Bay of Fundy today, the Triassic period began as a desolate time on Earth. Climate change and mass extinctions had marked the previous Permian period, and Pangaea's vast interior was hot and barren. A series of large rift valleys formed, cutting through the landscape as the supercontinent began to break apart. The Fundy basin was one of these.

Getting There

Driving Directions

From Route 1 east of Saint John, take Exit 137. Alternatively, from Route 1 in Sussex, take Exit 198. Follow Route 111 to its intersection (N45.34815 W65.55562) with Main Street in St. Martins. This point is about 40 kilometres from Exit 137 and 52 kilometres from Exit 196.

Follow Main Street east through St. Martins, around the harbour to the intersection (N45.35944 W65.53330) by the lighthouse. Fork right onto Big Salmon River Road and cross the covered bridge. Stay right, passing Alan Road, then follow Big Salmon Road as it curves left along the beach.

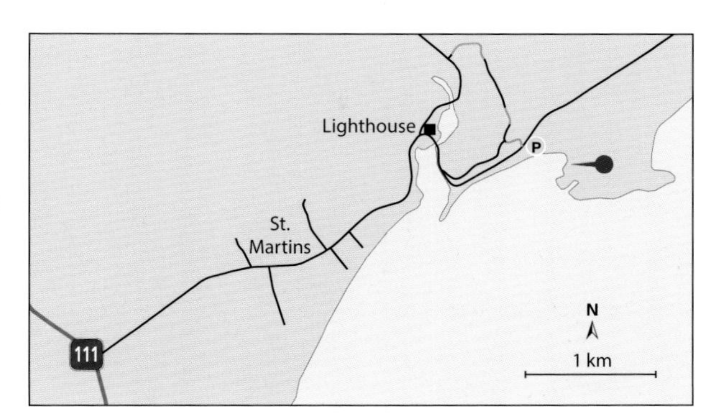

Where to Park

Parking Location: N45.35881 W65.52544

This is a gravel area near amenities on the east end of the beach.

Walking Directions

As conditions allow in low tide, walk east across the cobble beach toward the cliff and sea cave. A stream flows across the east end of the beach at low tide; hop or splash across if you can do so safely.

Caution

Fundy tides rise quickly. Consult tide tables for the day of your visit and plan carefully.

Notes

For more information about Stonehammer Geopark sites, visit www.stonehammergeopark.com.

1:50,000 Map

Loch Lomond 021H05

Provincial Scenic Route

Fundy Coastal Drive

On the Outcrop

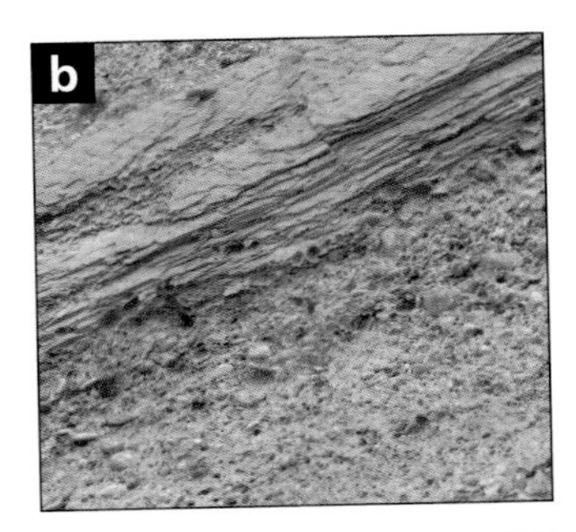

Along the unconformity (**a**), contrasting colours make the contact between (**b**) grey conglomerate and (**c**) red sandstone easy to recognize.

Outcrop Location: N45.35790 W65.52288

As you view the cliffs from the beach, the oldest layers are on the right. Each successive layer stacked above and to the left is a bit younger. The contact between (older) red sandstone and (younger) grey conglomerate is visible in a nook at the far end of the conglomerate cliff.

The red sandstone was deposited in an alluvial fan (see Exploring Further). Its coarse sand contains gravel derived mainly from volcanic rocks of the ancient Caledonia terrane. A few of the sandstone layers are cross-bedded, but many are not, which is typical of braided streams. Long, narrow streaks of light grey represent sediment that was starved of oxygen after burial, causing grey ferrous oxide to form.

The grey conglomerate, containing mainly quartzite cobbles, was deposited by a powerful braided river system flowing north across the Fundy basin. Heavy seasonal rains allowed the river to transport all this large rock debris far enough to round the cobbles. In the cliff face you can also see pods and discontinuous beds of coarse grey sandstone. These formed during periods of low water flow, allowing sand to settle in restricted channels.

Rock Unit	Bedrock Map
Quaco and Honeycomb Point Formations, Fundy Group	MP 2004-110

FYI

- Based on an analysis of the cobbles in the grey conglomerate, geologists can tell that they were deposited in water that was at least 2 to 3 metres deep and flowing quickly, in order to carry debris of that size. The rock's fine matrix probably settled and drifted down between the cobbles afterwards, during periods when less water flowed.

- The conglomerate in the cliff (photo **a**) is the source of most of the cobbles on the beach (photo **b**). On some of the cobbles you may see round or oval discoloured dents where cobbles have pressed against one another during burial and lithification of the conglomerate.

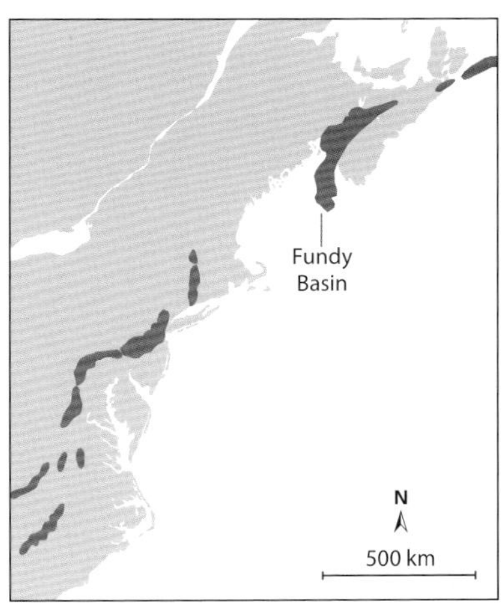

- Together they represent an early, but failed, phase of ocean opening. Finally, during the Jurassic period, the present-day Atlantic Ocean opened along a series of rifts farther east, now buried under sediments offshore.
- The Fundy basin took the form of a half-graben, with a steep fault scarp along the northwestern (New Brunswick) side.
- The Fundy basin is the largest of a dozen similar basins distributed along North America's eastern seaboard (see map at right).

Exploring Further

US National Aeronautics and Space Administration (NASA). "Badwater, Death Valley National Park." https://www.nasa.gov/image-feature/badwater-death-valley-national-park (aerial photo mosaic illustrating landforms similar to those of the Triassic Fundy basin, including alluvial fans and braided river channels).

Lookouts provide dramatic views, access to trails, and inviting facilities along the Fundy Trail Parkway.

Into the Highlands

Evidence of Avalonia's Dramatic Past

The Fundy Trail Parkway is an engineering marvel, winding along clifftops and across steep valleys near the rugged coast. Preparation of the route has exposed kilometres of rock face, giving geologists a detailed look at a region previously difficult to access.

This part of southern New Brunswick, known as the Caledonia Highlands, is mostly underlain by rock of the ancient microcontinent Avalonia. Folding and faulting has resulted in a complex pattern of outcrops along the parkway. But with the basic introduction provided here, you should be able to recognize many of the rocks that you see along your way.

The rock that dominates the parkway and nearby Walton Glen is rhyolite (outcrop 1) about 555 million years old. The rhyolite is so extensive, so thick, and so consistent in age that geologists think it is likely to have formed as the result of a single pyroclastic super-eruption. The parkway has also provided extensive new views of Cambrian sedimentary rocks (outcrop 2) and has revealed a sliver of one of the oldest known rocks in New Brunswick (outcrop 3).

Getting There

Driving Directions

Starting on Big Salmon River Road near the St. Martins sea caves (site 17), drive eastward about 8 kilometres, following signs to the park entrance (N45.38823 W65.45940).

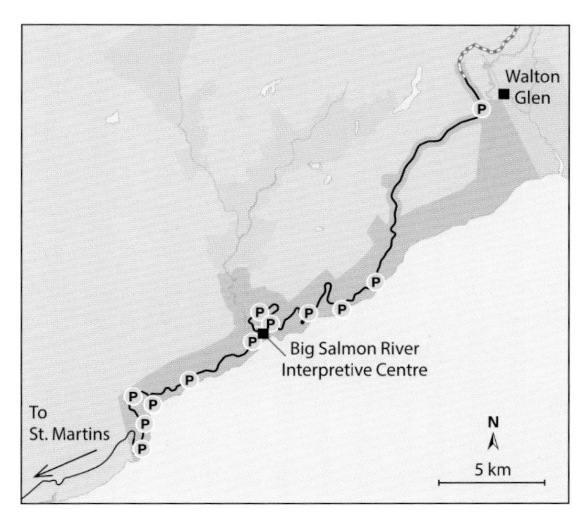

As of 2018, the parkway extends 30 kilometres from its western gate to Walton Glen. An extension planned for completion in 2021 (dashed line) will join the parkway to Route 114 and Fundy National Park.

The main rock units along the parkway can be viewed in the first 15 kilometres. The speed limit along the parkway is 30 to 40 kilometres per hour or less; and until the through-route is completed, you must double back to exit at the west end. Be sure to allow plenty of time for your round trip and be mindful of park closing times.

Where to Park

Parking Location: N45.42237 W65.40539

This location is in parking lot P8 below the Interpretive Centre at Big Salmon River, giving access to the suspension footbridge, a nearby lookout, and the Fundy Footpath. However, the parkway provides numerous other parking areas (see map) and small pull-offs.

Notes

For more information about the park and its amenities and programs, visit www.fundytrail.ca or ask for a brochure at the entrance.

1:50,000 Map

Salmon River 021H06

Provincial Scenic Route

Fundy Coastal Drive

On the Outcrop (1)

The first sizable outcrop along the parkway is colourful rhyolite.

Outcrop Location: N45.40213 W65.45146

As you travel from the west, this outcrop is about 500 metres beyond parking area P4. Typical of the less altered rhyolite rocks on the parkway, the colour scheme is pale pink to mauve; the layering and colour changes are vague. This rock type is quite brittle, and the outcrop surface is broken into sharp-edged, irregular fragments.

Many parkway outcrops, from here all the way to Walton Glen, formed from hot volcanic ash. While still hot, the ash fused together to form rocks known as welded tuff. Some varieties include crystal and/or rock fragments, while those that sagged or flowed while still hot are called flow-banded rhyolite.

To see these rocks in a natural setting, from the Big Salmon River Interpretive Centre, follow the trail along the river to the suspension footbridge (N45.42248 W65.40918). Outcrops along the south riverbank (orange dot on map, next page) and nearby stairway and lookout trail are rhyolite.

Volcanic rocks, Big Salmon River.

Rock Unit	**Bedrock Map**
Silver Hill Formation, Coldbrook Group	MP 2004-118 (rev. 2015)

On the Outcrop (2)

At this outcrop between parking areas P9 and P10, Cambrian sedimentary rocks have obvious layering in shades of grey and white.

Outcrop Location: N45.42878 W65.39808

East of the Big Salmon River, the road climbs and curves through high, tilted slabs of Cambrian sedimentary rock. Large-scale folds and faults have created a complex series of outcrops, and from here to Long Beach, the parkway passes in and out of these sedimentary layers (grey in the map at right).

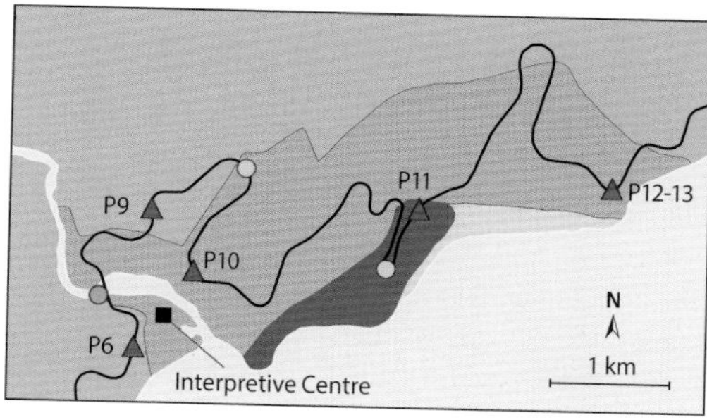

The outcrop pictured above is between parking areas P9 and P10, where the Cranberry Brook trail crosses the parkway. The layering is obvious, in some places with strong banding in shades of grey. In other outcrops you may see red-, purple-, or green-tinted shale or sandstone.

In some locations, the sedimentary rocks have been caught up in fault movements and crushed into zones of flaky fragments. Others have been stained dark brown, orange, or yellow by iron- and sulphur-bearing fluids in the fault zones.

Rock Unit

Saint John Group

Bedrock Map

MP 2004-118 (rev. 2015)

On the Outcrop (3)

Metamorphosed mafic rocks are well exposed along the far side of the parkway's Hairpin Turn.

Outcrop Location: N45.42493 W65.38721

For about 1 kilometre west of parking area P11, the parkway negotiates a tight bend known as the Hairpin Turn (see map for outcrop 2). The eastern side of the turn has no guardrails, and the roadbed widens a little and includes a grassy verge. Outcrops are visible within about 250 metres of parking area P11 and could be visited from there.

The mafic volcanic rocks here (dark green in the map on the previous page) are distinctly different from others along the parkway. Fresh surfaces are dark greenish grey (weathered areas are brown). Due to deformation and metamorphism, in some places the rock is riddled with white veins and pods of quartz.

Based on the age of a cross-cutting dyke, these rocks are at least 690 million years old. They formed during the Cryogenian period, a time of recurring, worldwide glaciation known as Snowball Earth. The rock's origin has been obscured by later events. It appears to have travelled here atop a Carboniferous thrust fault, so like a quotation out of context, its meaning is unclear.

Rock Unit	**Bedrock Map**
Long Beach Formation (unnamed group)	MP 2004-118 (rev. 2015)

FYI

- The rocks of the Caledonia Highlands vary in age from west to east. In the west they include Cambrian sedimentary rocks (dark grey in the map below) but are mainly volcanic rocks and related intrusions that are all about 560 to 550 million years old (dark and light pink respectively). The rocks in the east (including site 19, Point Wolfe) are also mainly volcanic and plutonic rocks (dark and light green respectively), but significantly older, about 625 to 615 million years. Both age groups were part of the microcontinent Avalonia.

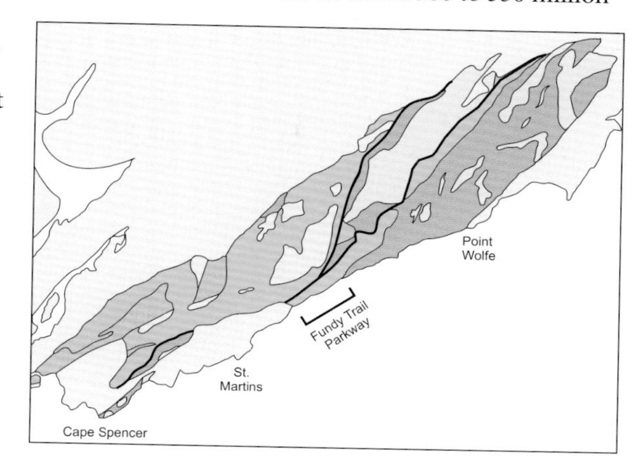

Also Nearby

Fundy Trail Parkway provides access to the trail into Walton Glen. This 60-metre-deep gorge is carved into a great thickness of rhyolite, part of the same rock formation seen beside Big Salmon River and in parkway roadcuts.

The Walton Glen parking area (P15; N45.491385 W65.310427) is about 10 kilometres northwest of Long Beach (P12/13). Outcrops along the parkway between Long Beach and Walton Glen are also mostly rhyolite.

Cliffs of Walton Glen Gorge.

Exploring Further

US Geological Survey. “What Is a Supervolcano?” https://www.usgs.gov/faqs/what-a-supervolcano? (include question mark in URL).

In Fundy National Park, the route to Point Wolfe beach includes one of New Brunswick's many covered bridges.

Explosive Past

Early Ediacaran Volcanic Rocks of Avalonia

Founded in 1948, Fundy National Park provides an area where natural environments unique to the Caledonia Highlands and Bay of Fundy's shores have been restored. Seventy years of protection have coaxed a thriving forest ecosystem from a site that had been cleared of its giant virgin trees and dotted with small-scale mining operations.

Picture the year 1910. At Point Wolfe was a small town beside a massive sawmill, its smokestack towering over the beach. Nearly a century of logging had decimated the forest. Nearby (accessible along the park's Coppermine Trail) a small mine had been operating for more than a decade. It was a busy, noisy site of industry.

Point Wolfe beach, now peaceful and reverting to its natural state, provides access to outcrops that imply an even noisier—and more noisome—environment. The volcanic rocks here formed about 620 million years ago in a series of sometimes explosive eruptions that were widespread across Avalonia. In fact, most of the park is underlain by volcanic rocks of similar age or by closely related intrusions.

Getting There

Driving Directions

On Route 114 by the Visitor Centre in the southeast corner of Fundy National Park (across the Upper Salmon River from Alma), watch for the intersection with Point Wolfe Road (N45.59528 W64.95060). Follow Point Wolfe Road for about 8 kilometres to its terminus at the Point Wolfe Campground (on the right). Parking is on the left.

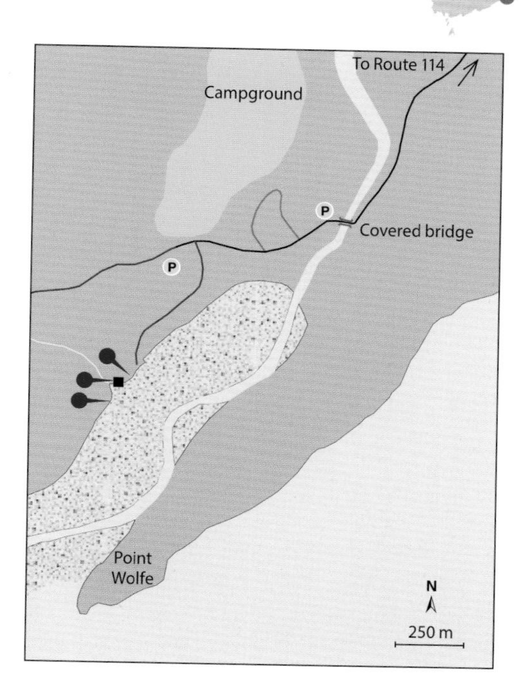

Where to Park

Parking Location: N45.54934 W65.01892

The parking area is located on the south side of Point Wolfe Road, across from the entrance to the campground. Alternatively, park near the covered bridge (N45.55088 W65.01372).

Walking Directions

From the parking location, follow signs for the trail to the beach. You may encounter a detour but will likely emerge along the shore in the vicinity of a large wooden structure (lookout and/or stairway) that may or may not be in use. West of the structure is a stream gully. Outcrops (along the rocky banks below the trees) are located east of this structure, between it and the gully, and beyond the gully.

Notes

This site lies within the boundaries of the Fundy National Park and requires a valid park pass.

Caution

Fundy tides rise quickly. Consult tide tables for the day of your visit and plan carefully.

1:50,000 Topo Map

Waterford 021H11

Provincial Scenic Route

Fundy Coastal Drive

On the Outcrop

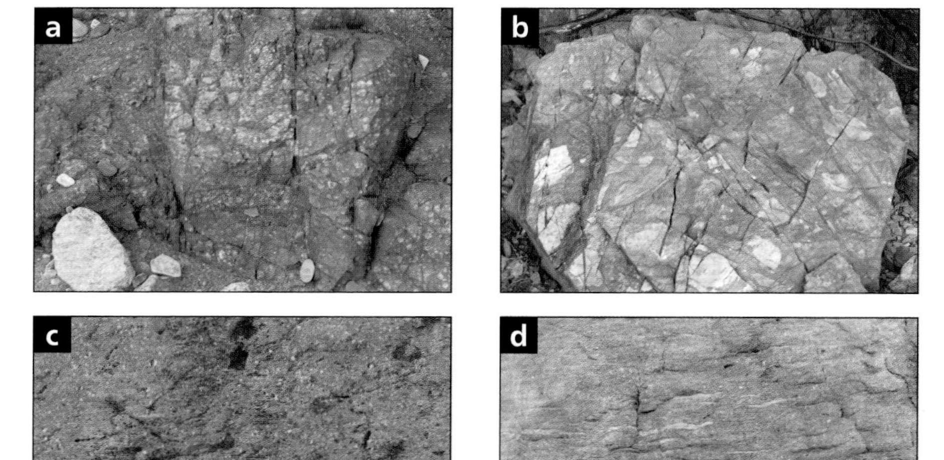

The variety of volcanic rocks along Point Wolfe beach include (**a, b**) lithic tuff and (**c, d**) crystal-lithic tuff with flattened lithic fragments.

Outcrop Location: N45.54671 W65.02074

At low tide, as you walk along the shore by the staircase and follow the estuary bank downstream, you'll see a showcase of volcanic rock types. Referred to as tuff, these rocks formed by the welding together of hot volcanic ash. Most of the tuff contains volcanic debris—rock fragments or visible crystals—in the fine-grained ash matrix.

The lithic tuff here (photos **a, b**) contains easily visible rock fragments measuring 1 to 10 centimetres or more across. The fragments were torn from the sides of a volcanic vent by the force of an explosive eruption. Most of the fragments in these tuffs are andesite, dacite, or rhyolite—rock compositions that are typical of explosively erupting volcanoes on continental margins above subduction zones.

The crystal-lithic tuff (photos **c, d**) is speckled with small white crystals of feldspar that formed as magma slowly cooled beneath a volcano. They were carried along as part of the ash cloud during an eruption. In some outcrops you may also see parallel strands of contrasting colour, about 2 to 5 centimetres long. This is volcanic debris pressed flat by the weight of accumulating ash layers.

Rock Unit

Broad River Group

Bedrock Map

MP 2004-124

FYI

- The early Ediacaran volcanic rocks in this region have been named the Broad River Group (green in the map). Along with related intrusions (light green in the map), they formed above a volcanic arc during subduction along a continental margin in Avalonia—as did the granite at site 16 (Cape Spencer). A few igneous rocks related to a 555-million-year-old super-eruption (site 18, Fundy Trail Parkway) are found in the northwest corner of the park (pink in the map); Carboniferous sedimentary rocks (shown in grey) occur along the coast in the southeast.

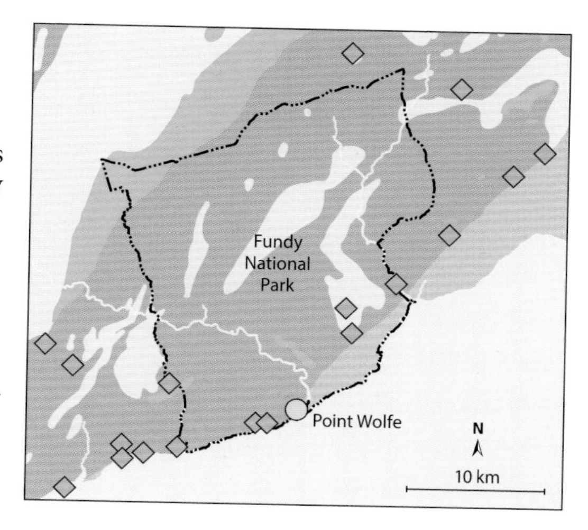

- The copper mine site near Point Wolfe beach is part of a larger pattern of copper mineralization in the area. In the geologic map above, reported finds of copper-bearing minerals are shown in orange. The presence of copper is limited at these sites, and none has proved economically significant. However, exploration outside the park is ongoing.

Also Nearby

- The tall reddish cliffs on the east bank of the estuary at Point Wolfe (photo at right) are Carboniferous sedimentary rocks similar to those at site 21, Hopewell Rocks. A fault along the river has moved them up against the ancient volcanic rocks on the west bank.

Carboniferous rocks, Point Wolfe.

- If you have a high-clearance vehicle, you can view a granite intrusion related to the volcanic rocks at Point Wolfe. Forty-Five Road follows the eastern boundary of the park. Along it, about 2.5 kilometres south of Old Shepody Road, a covered bridge (N45.68664 W64.95312) provides excellent views of the 620-million-year-old granite in nearby cliffs and in the Forty-Five River bed below.

The Cape Enrage lighthouse shares its dramatic view with a zipline, trails, and other amenities.

Formerly Mountains

River Sandstone of the Coal Age

At Cape Enrage, the Fundy tide rushes through a passage barely 9 kilometres wide. A rocky reef extends about 0.5 kilometres from the lighthouse—hidden at high tide but roiling the water into wild turbulence at other times. Hence the cape's name, and hence the need for a lighthouse, one of New Brunswick's oldest.

The rock at the cape is sandstone, part of a 1-kilometre-thick rock formation known for its pure, uniform sand. The sand was carried out of the young, rapidly eroding Appalachian mountain range as Gondwana's collision with Laurentia entered its final stages. Large, shifting river channels flowed through a broad valley—termed a braid plain—that included what is now southeastern New Brunswick and much of Nova Scotia's northern mainland.

Geologists have concluded that of all present-day examples, the Brahmaputra River, which flows out of the Himalayan mountains through India and Bangladesh, is most like the river system that deposited the sand in these layers at Cape Enrage. Consider the analogy and imagine the landscape: distant towering peaks, bordering foothills, and the tireless, hard-working river.

Getting There

Driving Directions

Access Route 915 from Route 114, for example near Alma (N45.60752 W64.93858) or in Riverside-Albert (N45.74314 W64.74029). From either direction, follow Route 915 south for about 17 kilometres to its intersection with Cape Enrage Road. Turn (N45.63536 W64.77816) south and follow Cape Enrage Road for about 6.5 kilometres to the entrance kiosk and parking location.

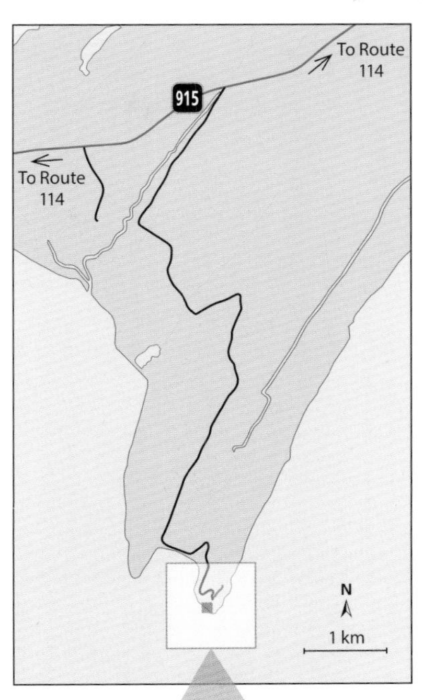

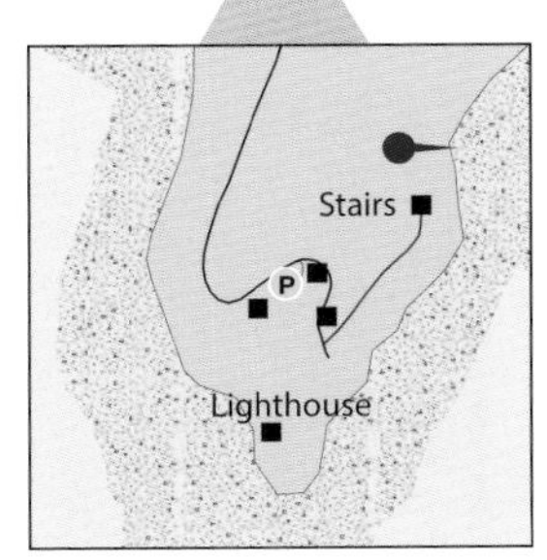

Where to Park

Parking Location: N45.59513 W64.77996

This is the lower of two large gravel parking areas.

Walking Directions

From the parking location follow the park road to a large gravel area near the lighthouse and zipline tower. From there, follow a gravel path northward along the shore about 150 metres to a large metal stairway. As conditions allow, walk down the stairway and onto the shore.

Notes

Rock exposures change over time as winter storms and other natural processes affect the rock cliffs. For advice on current features of interest, consult park staff at the Interpretive Centre. For more information about Cape Enrage hours, fees, and facilities, visit http://www.capeenrage.ca.

Caution

Fundy tides rise quickly. Consult tide tables for the day of your visit and plan carefully. In foggy weather, a loud horn may sound at this location.

1:50,000 Map

Alma 021H10

Provincial Scenic Route

Fundy Coastal Drive

On the Outcrop

People look like miniatures as they watch the ebbing tide beside high sandstone cliffs at Cape Enrage.

Outcrop Location: N45.59620 W64.77800

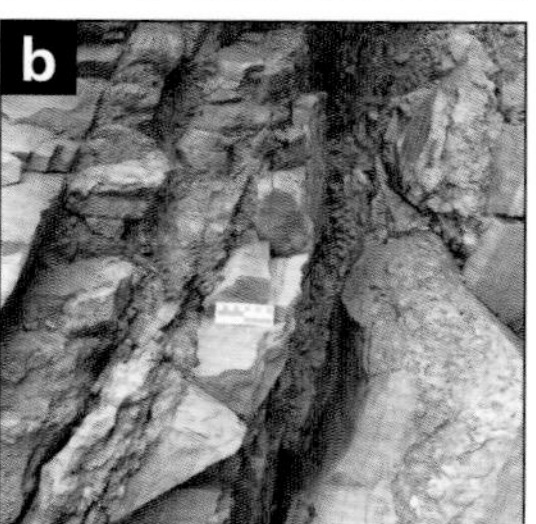

Deformation has fractured the sandstone so much it can be difficult to identify the layering as you walk along the base of the cliff. The beds stand nearly vertical and are roughly parallel to the cape's long eastern shore. (Check out a low-tide aerial view online to see the effect clearly.)

Most of the rock layers are uniform, fine- to medium-grained sandstone (photo **a**). These are river channel deposits. Now and then the river flooded, leaving lakes of standing water for a time. Such events produced a more complex series of layers, including pebble-rich varieties, fine silt- or mudstone, and small amounts of black shale or coal (photo **b**).

Fossils of plants and of animal tracks have been found at Cape Enrage. Enjoy any you see, but please don't remove them.

Rock Unit	Bedrock Map
Boss Point Formation, Cumberland Group	MP 2005-43

FYI

- This same rock formation in Nova Scotia has been quarried for more than a century as the Wallace sandstone. It was used to build the national parliament buildings in Ottawa and many other structures across eastern North America, including the Confederation Building in Charlottetown, Prince Edward Island (site 44).

Barn Marsh, seen here from the shore between the inner (left) and outer (right) ridges of Cape Enrage.

- Cape Enrage is two-pronged, like a seafood pick. The lighthouse stands on the outer cape; between it and the inner cape is a long, narrow wetland known as Barn Marsh. The rock of the inner cape is the oldest, a resistant, coarse red sandstone related to those found at Hopewell Rocks (site 21). Barn Marsh is underlain mainly by weak, easily eroded shale. The resistant sandstone of the outer cape is the youngest of the three rock units.

Also Nearby

Shepody Bay and its expansive marshlands attracted Acadian farmers in the early 1700s. Grindstones were essential for sharpening their scythes and other implements, and nearby Grindstone Island was a favourite location for extracting the stones—from the same rock formation seen at Cape Enrage.

For a view across Shepody Marsh to Grindstone Island, stop by Hopewell Hill Baptist Church (N45.76665 W64.68019), about 6 kilometres from Riverside-Albert or 11 kilometres from Hopewell Rocks on Route 114.

Shepody Marsh with Grindstone Island beyond.

Big Cove is one of several sites at Hopewell Rocks where visitors can walk among sea caves and towering sea stacks at low tide.

An Alluvial Rock Fan

An Alluvial Fan Deposit near a Major Fault

Hopewell Rocks is by far the most visited tourist attraction in New Brunswick. At low tide, you can walk on the exposed sea floor among the massive rock towers known as sea stacks—it's impressive. Because the lower, intertidal portions of the cliffs and stacks are obscured by seaweed, to see the rock features clearly, your best bet may be the trailside lookouts or a kayak tour at high tide.

The geological setting of these rocks was as dramatic as the present-day scenery. During the early part of the Carboniferous period, renewed uplift of the Caledonia Highlands formed a tall escarpment along a fault just a few kilometres inland from here. Streams rushed out of the high ground onto an adjacent plain; the sediment they had carried accumulated in a series of arc-shaped wedges known as alluvial fans.

These streams carried a jumbled load of rocky debris from the highlands but quickly lost energy in the flatter landscape of the plain. Boulders were deposited nearest the highlands, cobbles travelled farther, and so on (see FYI). Mainly pebbles and coarse sand got as far as this site.

Getting There

Driving Directions

From Route 114 about 13 kilometres south of Hillsborough, follow signs for Hopewell Rocks and watch for the park entrance (N45.82861 W64.58250). Follow Rocks Road into the park and fork right as indicated by signs. It's about 1.3 kilometres from Route 114 to the parking area.

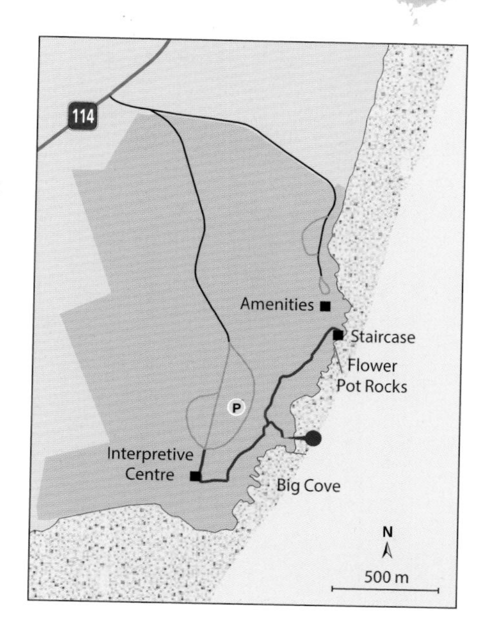

Where to Park

Parking Location: N45.81894 W64.57708

This large paved lot can be very full, especially around low-tide times.

Walking Directions

From the parking area, walk to the ticket area to gain entrance. From there, continue south to the playground near the gift shop and restaurant. Follow signs to Flower Pot Rocks. The main trail is about 750 metres long.

About 300 metres along the main trail, watch for the sign for Big Cove and walk about 100 metres to the lookout. (To visit Flower Pot Rocks, return to the main trail and continue to a pavilion and large metal stairway leading to the shore.)

Notes

A shuttle service to Flower Pot Rocks is available for those preferring not to hike the trail. For more information about the park, visit http://www.thehopewellrocks.ca.

Caution

Fundy tides rise quickly. Consult tide tables for the day of your visit and plan carefully.

1:50,000 Map

Hillsborough 021H15

Provincial Scenic Route

Fundy Coastal Drive

On the Outcrop

The cliffs at Big Cove lookout display roughly defined layers of cobbles, pebbles, and grit in a matrix of sand.

Outcrop Location: N45.81832 W64.57496

Big Cove lookout is one site among many in the park for a clear view of the rock features. Notice the tilted rock layers recognizable by a change in fragment size. Typical of this rock formation is a tendency for beds to "fine upward," that is, with fragments of decreasing size from the bottom to the top of a layer. This could reflect seasonal variations, just as now the spring freshet sends water rushing out of the highlands, while a dry summer may slow it to a trickle.

Immature sediment.

Because the sediment in these rocks is poorly sorted (sand, pebbles, and other sizes jumbled together) and many of the fragments are angular (rather than rounded), geologists label these sediments as immature. This is characteristic of deposits by streams that are choked with sediment from a nearby source.

The red colour of the outcrops is typical of sedimentary rocks formed in an arid or semi-arid climate.

Rock Unit

Hopewell Cape Formation, Mabou Group

Bedrock Map

MP 2005-48

FYI

- The proximal parts of alluvial fans are those nearest the highlands, while the distal parts are farthest away. Boulder conglomerate with fragments more than 1 metre across have been found on nearby Shepody Mountain, representing the fans' proximal deposits. Across Shepody Bay at Dorchester Cape, their distal deposits are mainly fine sand.
- The plain adjacent to the highlands had been previously occupied by the Windsor Sea (site 22, Hillsborough). The highlands were an earlier version of the present-day Caledonia Highlands (sites 18, Fundy Trail Parkway, and 19, Point Wolfe).
- Hopewell Rocks and those at Lepreau Falls (site 11) were formed around the same time in the same type of setting.

Also Nearby

The Albert County Museum (N45.84892 W64.57700; photo at right) is located along Route 114 about 2.5 kilometres north of the entrance to Hopewell Rocks Provincial Park.

The museum preserves a historic gaol (right in photo) made of stone from Grindstone Island. In the nearby Exhibition Hall (left in photo) are locally quarried grindstones as well as displays relating to the county's gypsum quarries and albertite mines (see site 22, Hillsborough). Albertite is a locally occurring, oil-rich shale, from which geologist Abraham Gesner derived the first supplies of kerosene in 1846. His discovery is recognized as the beginning of the oil and gas industry.

Exploring Further

Albert County Museum, https://www.albertcountymuseum.com.

Snair, Kevin. *Bay of Fundy's Hopewell Rocks.* Riverview: Chocolate River Publishing, 2015. A well-illustrated book that includes a self-guided tour, cultural history, and more.

Bright white outcrops of gypsum illuminate trailside clearings near the local golf course in Hillsborough.

C

Ancient Sea

Gypsum Deposits of the Windsor Sea

Acadian farmers who founded what is now Hillsborough in 1700 recognized the area's conspicuous white outcrops of gypsum as a valuable source of lime for their fields. Later, New England Planters processed agricultural lime for export. But from the mid-1800s to mid-1900s it was the demand for plaster-based construction materials that fuelled a massive gypsum quarrying, mining, and processing industry here.

This long-valued resource is the product of an ancient sea that once covered much of the Maritime region. For about 15 million years early in the Carboniferous period, sea level was unstable. It rose and fell numerous times, possibly due to the repeated formation and disappearance of extensive continental glaciers on Gondwana.

Seawater flooded into the low-lying Maritimes basin during times of high sea level. The resulting body of water has been named the Windsor Sea. Located at the equator, and with restricted connections to the open ocean, the Windsor Sea became saltier and saltier due to evaporation. Eventually becoming oversaturated with minerals, the sea deposited layers of first limestone, then gypsum, and eventually sodium and potassium salts.

Getting There

Driving Directions

On Route 114 across from the Hillsborough post office, find the intersection with Golf Club Road and turn (N45.92225 W64.64536) west. Follow Golf Club Road past the golf course clubhouse. About 450 metres beyond the clubhouse (N45.90401 W64.65683), fork right. Keep right as the road curves around the back of the course. Pass New Road on the left, then watch for a gravel track on the left (N45.89697 W64.66278) and turn in for the parking location.

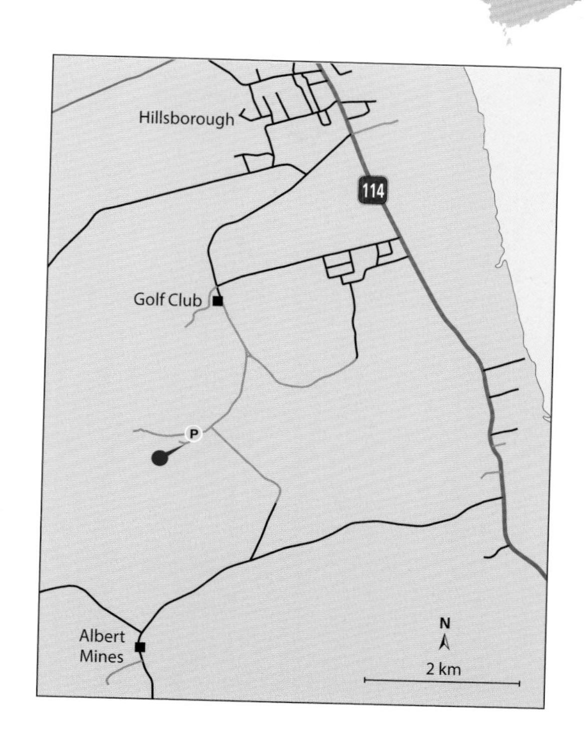

Where to Park

Parking Location: N45.89674 W64.66292

Park beside the gravel track so as not to impede other traffic.

Walking Directions

Follow the track through a wooded section and into an open meadow, where the outcrops are located.

Notes

This site is part of a network of tracks and trails popular with ATV enthusiasts.

1:50,000 Map

Hillsborough 021H15

Provincial Scenic Route

Fundy Coastal Drive

On the Outcrop

Weathering of gypsum causes many outcrops to crumble, but some are grooved or pitted by exposure to rain.

Outcrop Location: N45.89609 W64.66370

Much of the gypsum in this clearing takes the form of loose boulders and other rubble, probably related to past quarrying activity. Their surfaces have a fractured appearance, partly due to deformation of the layers and partly due to weathering. Look for areas of bright white and contrasting areas of bluish light grey—these are gypsum and anhydrite, respectively.

Gypsum is a calcium sulphate mineral that incorporates water into its crystal structure. Anhydrite is similar, but without the incorporated water. Each can be gradually converted to the other depending on how much water is available.

In a few places, exposed surfaces display a feature called rillenkarren (see photo), vertical grooves about 3 to 5 millimetres wide that look like claw marks. These are caused by rain running down the exposed surface (see FYI).

Rillenkarren on anhydrite.

Rock Unit

Upperton Formation, Windsor Group

Bedrock Map

MP 2005-48

FYI

- Rillenkarren form on water-soluble rocks such as limestone and gypsum. They form only on clean, unbroken, steep surfaces. Over a period of several decades, rain running down the rock surface slowly dissolves material to form the small channels.
- Sharp-rimmed pits on the top surface of some rocks at this site are also formed by exposure to rain. Rainwater trapped in the pit dissolves a micro-amount of rock, then a later rainfall splashes the material out.

Also Nearby

Albert Mines was arguably the birthplace of the petroleum industry. Between 1850 and 1880, its mines produced more than 200,000 tonnes of oil-rich shale (albertite) from which natural gas and kerosene were extracted.

Today a quiet rural community, Albert Mines is designated a Provincial Historic Site of New Brunswick to commemorate its geoheritage. To visit, from Route 114 about 4 kilometres south of Hillsborough, turn (N45.89212 W64.62669) west and follow Albert Mines Road about 3.7 kilometres to the cemetery, church, and former church hall (Gillen Hall). The mines were located about 2 kilometres to the west, along Old Albert Mines Road.

Exploring Further

Martin, Gwen. *Gesner's Dream: The Trials and Triumphs of Early Mining in New Brunswick*. Fredericton: Canadian Institute of Mining, Metallurgy, and Petroleum—New Brunswick Branch, 2003. A lively account of the region's mining history.

Stenson, Ronald, and Derek Ford. "Rillenkarren on Gypsum in Nova Scotia." *Géographie physique et Quaternaire*, vol. 47 (1993), pp. 239–243. https://doi.org/10.7202/032951ar (detailed study of rillenkarren formation in related rocks of the Maritimes region).

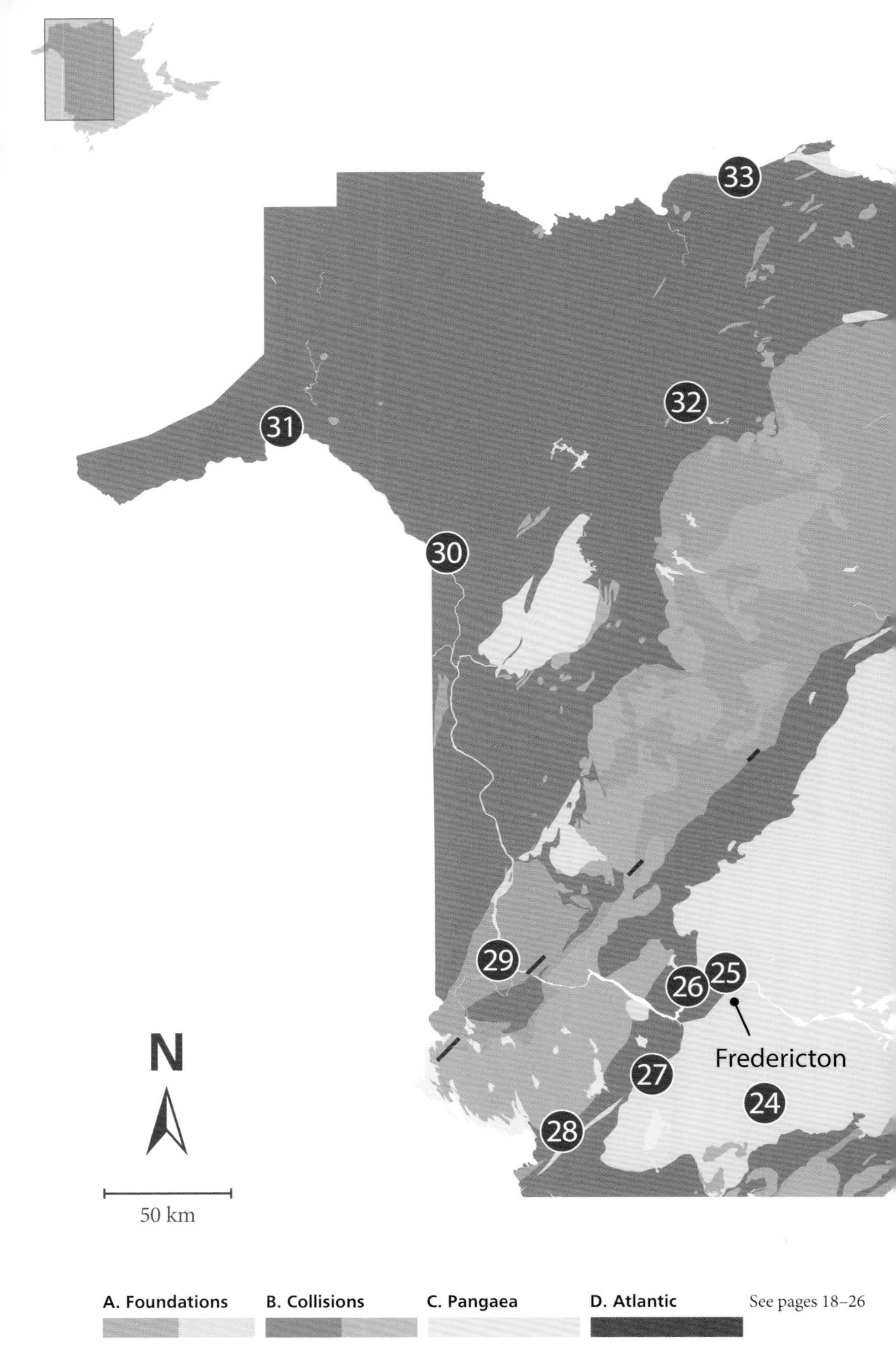

See pages 18–26

NORTHWEST

Tour 3 at a Glance

Tour 3 begins with a series of trips centred on New Brunswick's capital region, then follows Highway 2 north before crossing the high Appalachians to Chaleur Bay. Tectonic unrest is the underlying theme of this tour, but its story is bracketed by relative quiet. Its oldest rocks formed on a peaceful continental margin of ancient Ganderia and its youngest in a long-lived river environment on stable, solid land.

At these sites, you can explore ...

23 **Minto**	Centuries of coal mining around Grand Lake
24 **Fredericton Junction**	Coal-age conglomerate and sandstone
25 **Currie Mountain**	A Carboniferous mafic intrusion
26 **Mactaquac Dam**	Turbidites of the Fredericton trough
27 **Harvey**	Devonian volcanic rocks of the Maritimes basin
28 **McAdam**	Glacially transported granite on display
29 **Hays Falls**	Cambrian sediments of the Miramichi terrane
30 **Grand Falls**	Ordovician-Silurian rocks of the Matapédia basin
31 **Edmundston**	Devonian shale transformed into slate
32 **Williams Falls**	Signs of Devonian unrest
33 **Sugarloaf**	Conduit beneath a Devonian volcano

Reforestation and the efforts of local mountain-bike enthusiasts have transformed several of Minto's strip-mined tracts into inviting recreational areas, like the Coal Mine trail shown here.

Coal Town

Centuries of Coal Mining around Grand Lake

Acadian blacksmiths needed it. American colonists sailed up the coast for it. The industrial revolution depended on it. From 1639 to 2010, coal brought people to this region at the head of Grand Lake. Retired miners and the descendants of miners still reside in Minto, keeping nearly four centuries of mining heritage alive.

The coal mined here was a single, nearly horizontal seam about 45 centimetres thick. Around Minto it lay near the surface under a layer of shale. For strip mining, 34 million cubic metres of shale were moved aside for every tonne of coal produced. The resulting landscape, known as "cut and dump," is still evident in several tracts in the area. The one pictured above, mined and reforested in the 1940s, is now a mature woodland—trees seem to like the shale.

The story of Minto coal began in the Carboniferous period just over 300 million years ago. A swampy forest of primitive plants included tree-sized relatives of today's ferns and mosses. Plant remains accumulated over tens of thousands of years before being buried and converted to coal.

Getting There

Driving Directions

Trails site (1): From Route 10 about 10.5 kilometres east of Little River, fork left (N46.05958 W66.09205) onto Northfield Avenue. Watch for signs for MTB Trail about 200 metres from the intersection. **Mural and museum (2):** Continue northeast along Northfield Avenue about 2 kilometres to Main Street. Turn right (N46.07667 W66.07600) and continue about 180 metres to a parking location near the Foodland and museum. **Library (3):** Continue along Main Street about 1 kilometre to the intersection with Pleasant Drive (Route 10; N46.07149 W66.06298). The building is on the northeast corner.

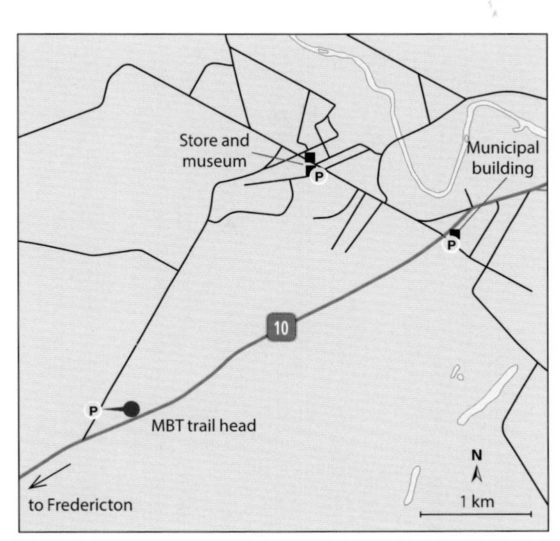

Where to Park

Parking Locations: Mountain-bike trails (1): N46.06144 W66.09058
Mural and museum (2): N46.07570 W66.07400
Library (3): N46.07134 W66.06241

(**1**) Park along the roadside, being careful not to block traffic. (**2**) Park in the lot by Veteran's Park or other nearby space. (**3**) Use parking spaces for library visitors at the municipal building.

Walking Directions

Trail site (1): Follow signs to the kiosk at the trail head—maps of the trails may be available there. Follow the Beginner or Coal Mine trail, staying alert for bike traffic. **Mural and museum (2):** Enter the Foodland to view the mural; from the store, cross Main Street to explore the museum and its grounds. **Municipal building/ library (3):** Enter from the south side; mount one flight of stairs for the library.

1:50,000 Map

Minto 021J01

Provincial Scenic Route

River Valley Scenic Drive

On the Outcrop

(**a**) Trail footbridge between ridges of shale "dump"; (**b**) dragline at work, from the Minto mural by Ron Sajack; (**c**) and (**d**) plant fossils found in the local mines.

Outcrop Location: N46.06144 W66.09058

A self-guided tour of Minto's mining heritage begins along Northfield Avenue (waypoint above) for access to walkable portions of local mountain-bike trails (photo **a**). From the trails you can get a feeling for the scale of the older strip-mining operations and for the masses of soft shale that were moved aside.

One local tribute to Minto's past includes a portrait of Maid Marion (photo **b**), the last and largest strip-mining dragline in operation near Minto. It is part of a mining-themed mural commissioned by the owner of the local Foodland grocery store and is on display inside. Across Main Street is the Minto Museum, with displays of photos, documents, equipment, and scale models related to mine operations.

Minto village emblem.

The town's mining-themed flag flies outside the Municipal Building. Inside, display cases in the hallway and in the library contain objects and photos related to the town's mining history, including plant fragments from the ancient coal forest, such as Stigmaria (a root from a giant, treelike club-moss, photo **c**) and fern foliage (probably Neuropteris, photo **d**).

Rock Unit

Minto Formation, Pictou Group

Bedrock Map

MP 2006-2

FYI

- Coal forms from peat, partially decayed plant matter that accumulates in wet conditions. For the Minto seam, geologists have analyzed plant fossils, fossil spores, and properties of the coal itself in order to find out more about its origin.

  The Minto coal seam appears to have originated in a poorly drained river delta in a subtropical climate, providing favourable conditions for plant growth and peat formation. Plants were mostly lycopsids (tree- and shrub-sized relatives of modern clubmoss) and sphenopsids (relatives of modern horsetails), with lesser amounts of ferns.

- At present tropical rates of peat formation, it would take about 2,000 years to form 5 metres of peat. That's enough to make about 0.5 metres of coal, about the thickness of the Minto seam.

Also Nearby

The Minto Museum is located in the town's former railway station, closely linked to the history of coal.

On the grounds of the Minto Museum are two commemorative plaques. One, placed in 1930, honours Minto as the site of North America's first export of coal (photo below). The other, dating from 1982, commemorates the accidental deaths in 1932 of two heroic miners and the three boys they sought to rescue from an open mine shaft.

Exploring Further

Ross, K. [kate ross]. *1954, underground mine.* https://vimeo.com/115285203 (vintage movie footage of underground mining operations in Minto, attributed to Aubrey Wasson).

The Oromocto River's North Branch flows between rocky banks in Peterson Rock Park, Fredericton Junction.

C

Rivers Then and Now

Coal-Age Conglomerate and Sandstone

Fredericton Junction began as the site of a water-powered mill established in 1804 by settler Thomas Hartt along the Oromocto River. In the next century, rail lines following the north and south branches of the Oromocto intersected here, making the community a busy transportation hub. Rivers have played a central role in the town's history—and in its geology, too.

Geologist Abraham Gesner explored along the North Branch in 1837. "At Mr. Hartt's mills … the banks of the north branch of the Oromocto are composed altogether of sandstones, shale, and conglomerate … all belonging to the great Coal District of the Province," he reported. The rocks here belong to the same series of layers as those at Minto (site 23), where abundant coal was mined for centuries.

As the Carboniferous period waned, the Maritimes basin broadened and gradually subsided. Through the basin flowed powerful, meandering rivers carrying silt, sand, and gravel from sources deep within the adjacent Appalachian mountain range. In Fredericton Junction you can view rocks formed from sediments deposited by ancient predecessors of the present-day Oromocto River.

Getting There

Driving Directions

From Route 101 in Fredericton Junction, about 100 metres south of the Oromocto River bridge, watch for Mill Pond Drive (N45.65613 W66.60676) and turn west. Follow Mill Pond Drive about 150 metres and turn right into Peterson Rock Park.

From the park, continue north on Mill Pond Drive for about 300 metres. It becomes Currie Lane and ends in a cul-de-sac at Currie House Museum (N45.65942 W66.61056).

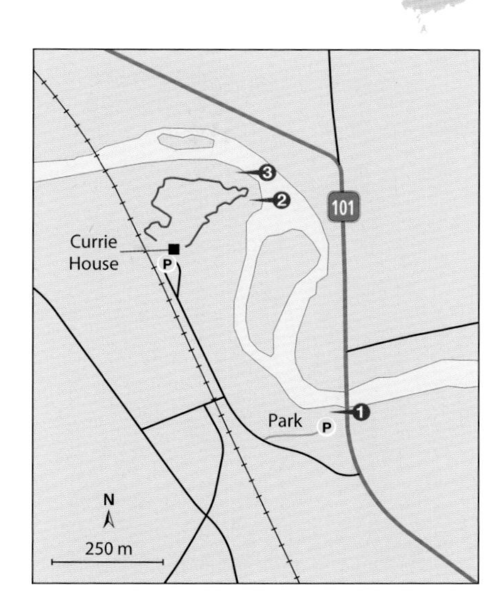

Where to Park

Parking Locations: Park: N45.65678 W66.60747
Museum: N45.65926 W66.61081

At Peterson Rock Park, use the gravel area at the end of the park lane. At Currie House Museum, pull well off the cul-de-sac, leaving room for other traffic.

Walking Directions

In Peterson Rock Park, as conditions allow, walk toward the riverbank near the road bridge. With moderate to low water levels, a rough rock pavement is exposed along the bank (1). At Currie House Museum, follow a gravel path along the left (west) side of the house to the trail head by the edge of the woods. The 600-metre-long trail is well marked; follow it to the lookout (2) above the rapids. About 40 metres west of the lookout, a narrow footpath leads down onto the riverbank. As conditions allow, descend onto the rocky pavement beside the river (3).

Notes

These sites are part of the Oromocto River Watershed Association's White Rapids Adventure Trail, http://www.oromoctowatershed.ca/location/.

1:50,000 Map

Fredericton Junction 021G10

Provincial Scenic Route

River Valley Scenic Drive

On the Outcrop

(**a**) In Peterson Rock Park, pebble-rich conglomerate lines the riverbank. (**b**) The cliff overlooking the rapids near Currie House is formed of sandstone.

Outcrop Location: N45.65688 W66.60755

You may think that this pebble conglomerate along the riverbank looks like badly mixed concrete. But it's a fairly typical river channel deposit: slightly rounded rock fragments of many sizes up to 4 centimetres across mixed with coarse sand (photo **c**). Such deposits are sometimes called lag gravels, because any fine particles travelling with them were washed away by the rushing water of the channel.

In contrast, at the lookout on the trail at Currie House (N45.66013 W66.60911), the rock outcrop is composed of well-sorted sand (photo **b**). Along some vertical surfaces you may see subtle cross-bedding. Some sandstone outcrops along the river (N45.66064 W66.60943) preserve ripple marks (photo **d**) formed by flowing river water.

Rock Unit

Minto Formation, Pictou Group

Bedrock Map

MP 2005-33

FYI

- Large or small, many of the world's rivers meander, following complex, curved paths when flowing across areas that don't constrain their banks. Near Peterson Rock Park and Currie House, rocky outcrops prevent the Oromocto riverbed from shifting easily. But a few kilometres downstream from Fredericton Junction (FJ in the map, **a**), where the north and south branches flow together, the river has meandered freely. The mighty Amazon River has developed a similar pattern (photo **b**).

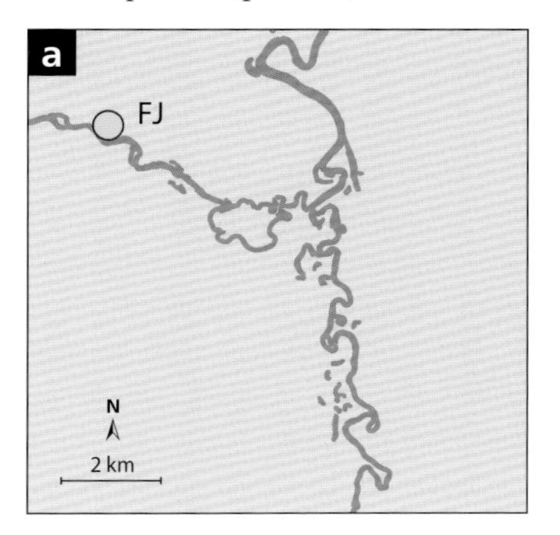

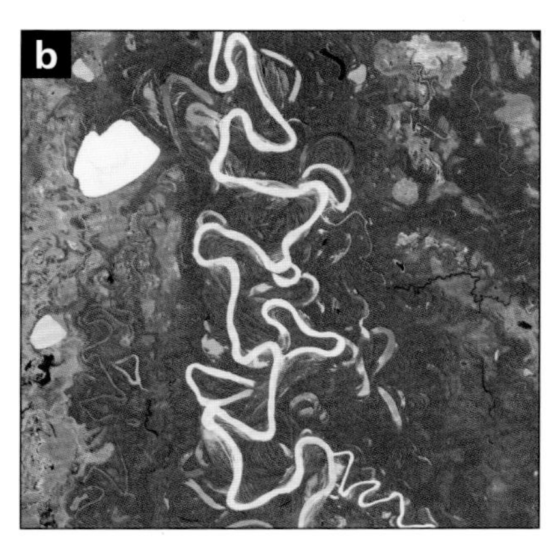

- In a meandering river, water flows more quickly around the outside of the curve, eroding that surface while depositing sediment on the inside of the curve. In this way meanders can change their curvature and location over time, covering a broad landscape with river sediment: gravel (conglomerate) in the channels, cross-bedded sand (sandstone) in the point bars, and mud (shale) in the flood plain during high water.
- Although the rocks of Fredericton Junction belong to the same sequence of layers as those around Minto (site 23), at this site local conditions did not favour coal formation. Nevertheless, any plant fossils you may see here are of similar age to those found in the Minto coal seam.

Exploring Further

Gesner, Abraham. *First Report on the Geological Survey of the Province of New-Brunswick*. Saint John: H. Chubb, 1839. http://online.canadiana.ca/view/oocihm.44810 (reports for 1840–1842 are at related URLs …/oocihm.44811, …/oocihm.44812, and …/oocihm.44813).

NASA Earth Observatory. "Meandering in the Amazon." https://earthobservatory.nasa.gov/images/84833/meandering-in-the-amazon (pull the dividing line back and forth to view changes to meanders in the Amazon River over a 30-year period).

A steep wall of gabbro borders Route 105 near a pull-off along the New Brunswick Trail west of Fredericton.

Little Volcano

A Carboniferous Mafic Intrusion

Based on historical accounts, Currie Mountain appears to have been named for Lieutenant Ross Currie, a Pennsylvania loyalist who in the 1780s settled nearby in what was then known as Douglas (now part of Fredericton). The hill is a conspicuous feature, rising to 80 metres beside Route 105 and the adjacent New Brunswick Trail along the bank of the Saint John River.

The hill is not only a topographic anomaly—it's a geological one, too. The rock here is a fine-grained gabbro of Carboniferous age. It forms a small intrusion into the much more widespread sedimentary rocks of the Mabou Group. Signs of related volcanic eruptions appear at a nearby site, so it's likely that a small volcano (long since removed by erosion) lay some distance above what is now Currie Mountain. Picture it puffing away in a landscape of rivers, sand, and primitive vegetation.

By this point in geologic time, Ganderia, Avalonia, and other terranes that had split from Gondwana were cemented to Laurentia; that era of collision and orogeny was complete. Now tectonic unrest took a different form—wrenching movements, tearing at the newly consolidated crust.

Getting There

Driving Directions

On Route 105 west of Fredericton, about 1.8 kilometres west of the intersection with Carlisle Road, watch for a small grass clearing (N45.98172 W66.75365) on the wooded north side of the road. This is the trail head.

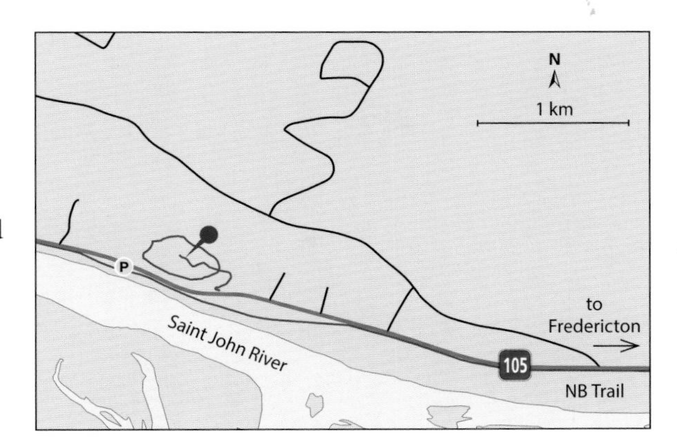

Where to Park

Parking Location: N45.98270 W66.75951

This large pull-off is located beside the New Brunswick Trail on the south side of Route 105. It is across the road from the prominent cliff face depicted on the previous page. Some hikers prefer to park near the trail head if conditions allow.

Walking Directions

From the clearing at the trail head, follow the trail on the left. (A chain across it discourages wheeled traffic.) It makes a small switchback up the initial slope. The southern half of the circular trail and the summit trail have the most outcrops. Near their intersection you'll find the remains of a brick chimney or fireplace.

Notes

The trails are minimally maintained but normally are recognizable due to frequent use.

1:50,000 Map

Fredericton 021G15

Provincial Scenic Route

River Valley Scenic Drive

On the Outcrop

Outcrops and boulders of gabbro cap the shady summit of Currie Mountain.

Outcrop Location: N45.98307 W66.75580

Exposures of the gabbro are scarce along the lower part of the trail, but boulders are common along the southern leg of the circular trail. Near the summit are numerous outcrops among the trees.

Most exposed surfaces have been weathered to shades of brown, and many are lichen covered. If you find a recently broken fresh surface, you'll be able to see the true charcoal grey to black colour of the rock. The rock typically contains visible crystals of feldspar up to 1 millimetre long in a finer-grained matrix. Pick up a loose piece of the rock along the way—you'll find it seems unusually heavy for its size. The magnesium- and iron-rich composition of gabbro makes it noticeably denser than a typical granitic or sedimentary rock.

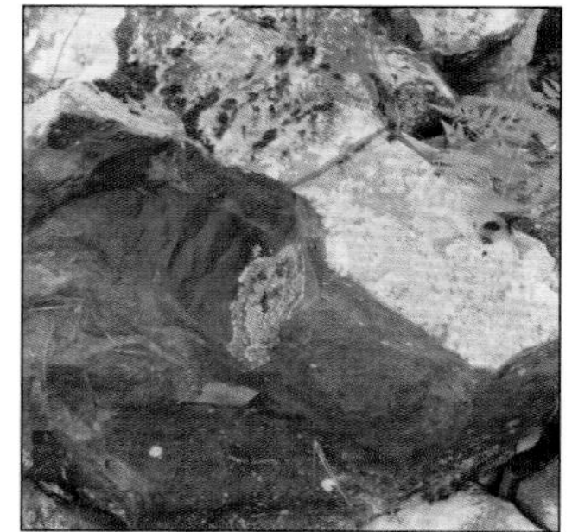

Broken fresh surface.

The large number of boulders on the hill may be due to the gabbro's columnar jointing. The jointing is not apparent along the trails but is visible in the roadside outcrop (photo, p. 144).

Rock Unit

Royal Road Basalt

Bedrock Map

MP 2005-38

FYI

- Nearby quarries on Carlisle Road and Royal Road expose lava flows similar in mineral and chemical composition to the rocks of Currie Mountain. Features in the lava layers suggest they flowed from the direction of Currie Mountain, meaning its rocks likely formed in a "feeder" channelling magma to a small volcano.

  In the quarries, the tops of some lava flows show signs of having become weathered prior to the next flow. Some flows even have layers of sandstone between them. These are signs that the volcanic activity was intermittent.

- When strike-slip fault motion causes crustal blocks to move side by side, space can open along faults with irregular boundaries. This process is known as transtension (in the diagram, as seen from above). As happened with the intrusion at Currie Mountain, magma may rise into such spaces.

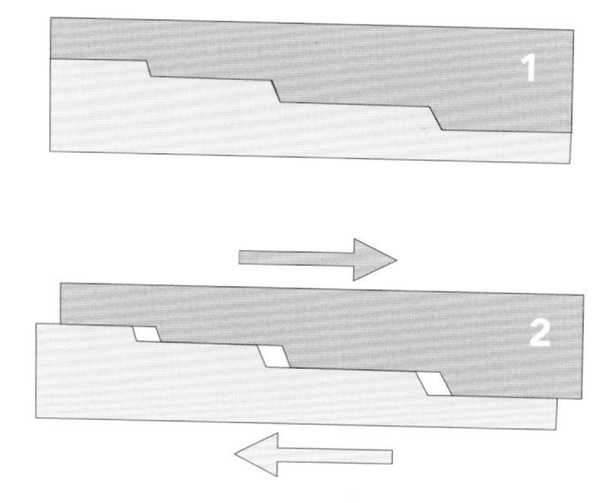

- Late in the Devonian period and early in the Carboniferous period this process affected numerous sites scattered throughout what is now southern and central New Brunswick (map at right). At all these sites, volcanic rocks or small intrusions occur within sequences of river-deposited sandstone. These include rocks near site 3 (on Bar Road across from Ministers Island) and at site 13 (Taylors Island).

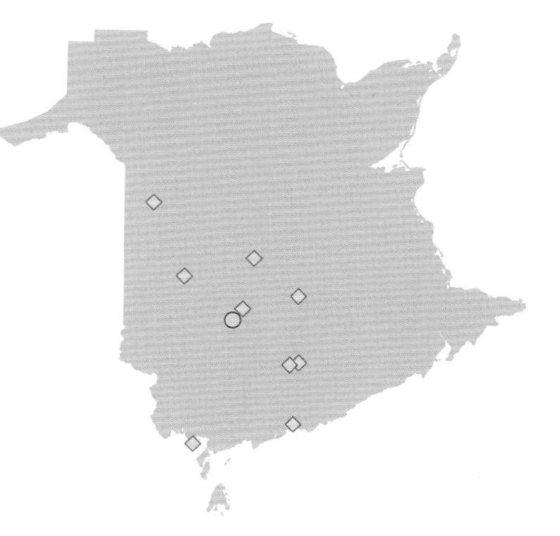

Rock outcrops near the Mactaquac Dam Visitor Centre are visible beside the parking area and nearby lane.

Rushing Cascades

Turbidites of the Fredericton Trough

The generating station at Mactaquac Dam, in service since 1968, is New Brunswick's largest hydroelectric facility and one of five on the Saint John River. The dam holds back the river to a depth of 40 metres, controlling and funnelling the water flow past the generators' giant turbines. In times of high water, excess flow tumbles down the dam's spillways.

Listen with your imagination and you might hear an ancient rush and hiss of tumbling currents, too. Rocks in the outcrops at this site were formed by turbidity currents—sediment-laden underwater cascades—that rushed down the sides of a deep marine basin known as the Fredericton trough. Most geologists agree that the Fredericton trough formed part of the last remaining seaway between Ganderia and Laurentia. The seaway filled and narrowed as the two continental blocks approached one another during the Silurian period.

Sediment from Ganderia entered the basin from what is now the east, while sediment from Laurentia entered from what is now the west. Deformation was inevitable during the resulting collision, but fortunately some surviving rock features can still tell us their story.

Getting There

Driving Directions

Follow Route 105 to the vicinity of Mactaquac Provincial Park. On the east side of the river, about 150 metres east of the sharp elbow turn, watch for a large sign at the entrance to Mactaquac hydro station at Power House Lane. Turn (N45.95974 W66.87093) and follow the lane downhill to the parking area by the Visitor Centre.

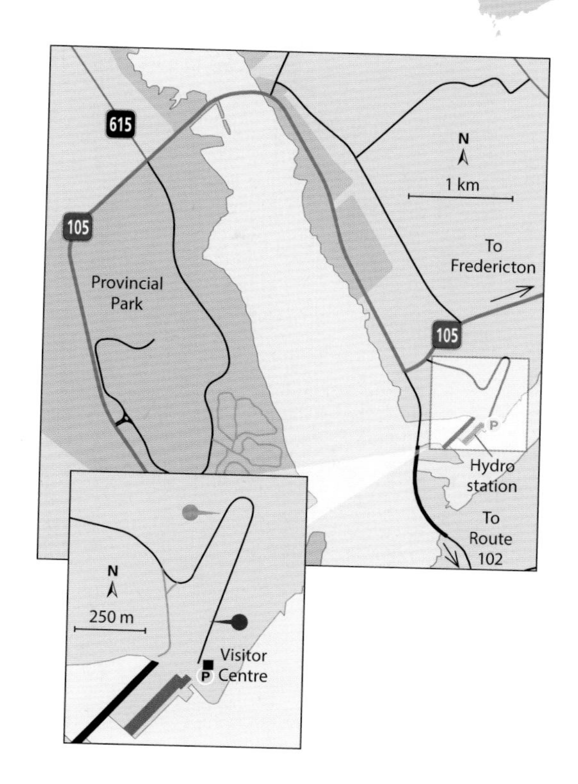

Where to Park

Parking Location: N45.95585 W66.86478

Park in the large paved lot between the Visitor Centre and the hydro station.

Walking Directions

From the Visitor Centre, walk back up Power House Lane about 150 metres to a rock face on the west side of the road.

Caution

Although the access road to the dam site typically gets little traffic, please use caution in approaching and examining the outcrop.

1:50,000 Map

Fredericton 021G15

Provincial Scenic Route

River Valley Scenic Drive

On the Outcrop

In this roadside outcrop near the dam's Visitor Centre, turbidite layers are steeply tilted, revealing the undersides of the original bedding.

Outcrop Location: N45.95709 W66.86481

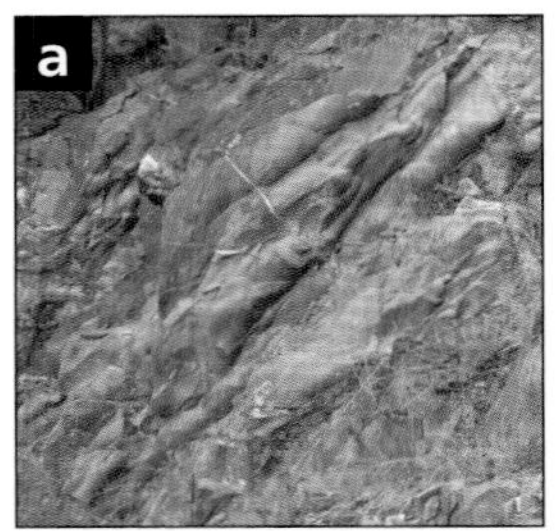

A feature called flute casts makes this outcrop particularly interesting. Flute casts are found on the underside of some turbidite beds—a surface conveniently exposed here (photo above, and photo **a** at right) because the rocks have been tilted by deformation. Flute casts look like long, curving welts on the rock surface. In this outcrop they are about 10 centimetres wide on average and occur in bulbous sections 30 or more centimetres long.

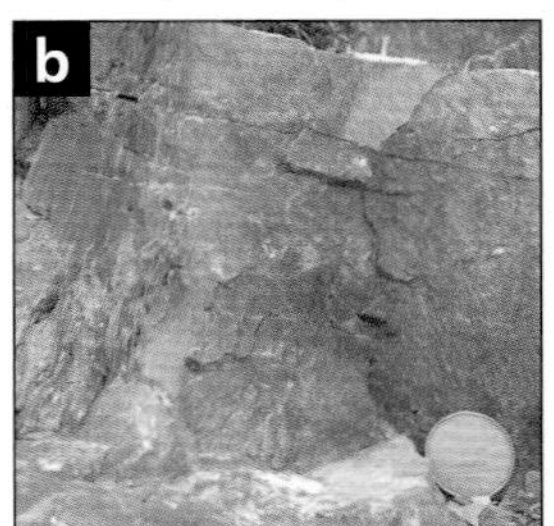

Another feature here, graded bedding, is also typical of turbidites. It is best seen on a broken surface that cuts across the bedding. Looking at photo **b**, for example, the coarse, sandy portion of the bed is on the right (above the loonie) and is nearer the road. The finer-grained shale portion, seen on the left, is farther from the road, toward the hillside.

Rock Unit

Burtts Corner Formation, Kingsclear Group

Bedrock Map

MP 2005-38

FYI

- Turbidites are formed by turbidity currents, fast-moving underwater avalanches of water and sediment. Each cycle of coarse-to-fine sediment represents the passage of a single turbidity current.
- As a turbidity current races down a slope, the swirling water scoops out hollows and dents in the surface it travels across (usually fine-grained sediment from a previous turbidity current). New, coarse sediment then settles out of the water and fills in those hollows. The sculpted pattern remains on the base or sole of the new bed, resulting in a type of sole mark known as flute casts. They capture the very moment at which the swirling current passed by.
- Turbidites can also be found on Campobello Island (site 6, Herring Cove). They are about the same age as these, but formed in a different sedimentary basin.

Also Nearby

Uphill from the hairpin bend along Power House Lane, nearly horizontal, reddish outcrops (N45.95957 W66.86460; photo above) are conspicuous beside the lane. These layers formed during the Carboniferous period and are lying above the older, Silurian turbidites seen farther downhill near the Visitor Centre.

Exploring Further

Institute of Earth Physics of Paris, Department of Lithosphere Tectonics and Mechanics. "Simulation of Turbidites." https://www.youtube.com/watch?v=CE4vdar8-NA (video of an experimentally produced turbidity current, with views from the side and from above).

Cherry Mountain, as seen from the entrance to Harvey Lakeside Trail, is a notable high point in the region's topography.

Fiery Cloud

Devonian Volcanic Rocks of the Maritimes Basin

Harvey Lakeside Trail loops through a village nature park on the flanks of Cherry Mountain beside Harvey Lake. Part of the trail passes through a meadow where a large lumber operation was once located. Part of it traverses adjacent wooded areas and includes a lookout over the lake.

Cherry Mountain is part of a string of hills running northeast-southwest through Harvey—about 20 kilometres long in all. The hills are mostly made of felsic volcanic rock, which is not to say the hills themselves were once volcanoes. Instead, it's likely this volcanic material was produced by the explosive eruption of a volcano about 35 kilometres to the southeast (see FYI).

By the end of the Devonian period, this part of the Earth's crust was being stressed and distorted by tectonic collisions far away to the south. Locally, heat flowing up from the mantle caused part of the lower crust to melt, forming sticky, silica-rich magma that rose toward the surface, eventually resulting in the catastrophic eruption. Along the trail you can see the rock formed by this dramatic event.

Getting There

Driving Directions

The village of Harvey is located along Route 3, about 18 kilometres south of Route 2 or about 14 kilometres north of the intersection with Route 4. Near the town centre, Route 3 curves as it crosses a railway line. Between the Canada Post building and the railway tracks, turn (N45.72954 W67.00722) northwest onto Cedar Lane. Continue along the lane about 140 metres to a gravel area beside a boulder-decorated garden at the entrance to Harvey Lakeside Trail.

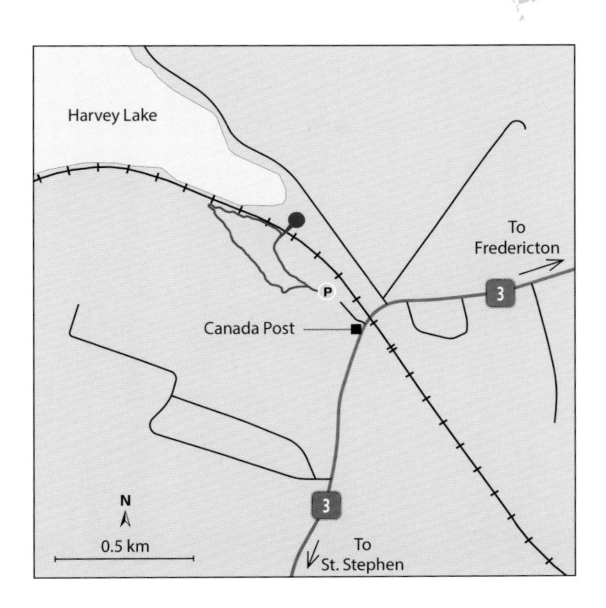

Where to Park

Parking Location: N45.73043 W67.00846

Park in the gravel area beside the boulder-decorated garden at the entrance to the trail grounds.

Walking Directions

Rather than walking uphill toward the kiosk, follow the eastern portion of the trail, which runs along the length of the field. As the trail turns, but before it enters the woods, look for a low, brick-coloured, jagged outcrop on the right-hand side of the trail.

Notes

You may see a map of the trail on a large sign near the parking area. On high ground overlooking the entrance is a picnic kiosk for visitors' use.

1:50,000 Map

McAdam 021G11

Provincial Scenic Route

River Valley Scenic Drive

On the Outcrop

A low, jagged outcrop of felsic crystal tuff is located immediately beside Harvey Lakeside Trail, near where the trail enters the lakeside woods.

Outcrop Location: N45.73154 W67.01005

This modest outcrop offers several fresh surfaces on which you can see the rock texture quite well. The rock has a speckled appearance due to many visible crystal fragments 1 to 2 millimetres across. The white fragments are feldspar and most of the darker, almost black, fragments are a variety of quartz. You may also notice a few small, perhaps red, irregular shapes. These are bits of felsic volcanic rock such as banded rhyolite.

Surrounding the crystal fragments is a brick-red or purplish matrix of welded volcanic ash (see FYI). The rock has the physical properties of very strong brick or porcelain, resisting breakage except where pre-existing cracks have weakened it. As the colour suggests, it is rhyolite (that is, akin to granite in composition).

Volcanic rock with a matrix of ash is known as tuff. Because this rock contains identifiable crystal fragments, it is called a crystal tuff.

Crystal tuff.

Rock Unit

Cherry Mountain Formation, Harvey Group

Bedrock Map

MP 2005-34

FYI

- Ash flow is a term geologists use to describe an aerosol of magma that explodes from a volcano in a very hot (up to 1000°C), dense cloud. This cloud of magma and volcanic gas roils and rushes over the landscape at 100 kilometres per hour or more. It deposits still-hot material, which then welds together under its own heat and weight. This phenomenon has other names, including nuée ardente (glowing cloud) or pyroclastic flow.
- The rocks of Cherry Mountain are unusual in some aspects of their chemistry. For example, they contain more silica, potassium, fluorine, boron, and uranium than a typical granite or rhyolite would. This pattern is typical of felsic melts formed very high in a magma chamber, where volatile elements are concentrated.
- The volcanic rocks around Harvey, while unusual, are not completely alone. Just 35 kilometres to the south-southeast around Piskahegan and Mount Pleasant is another group of volcanic rocks of the same age. The rocks at that site indicate that a large volcano erupted catastrophically and collapsed, forming a caldera. Geologists think that ash flows from that eruption were the source of the volcanic rocks at Harvey.

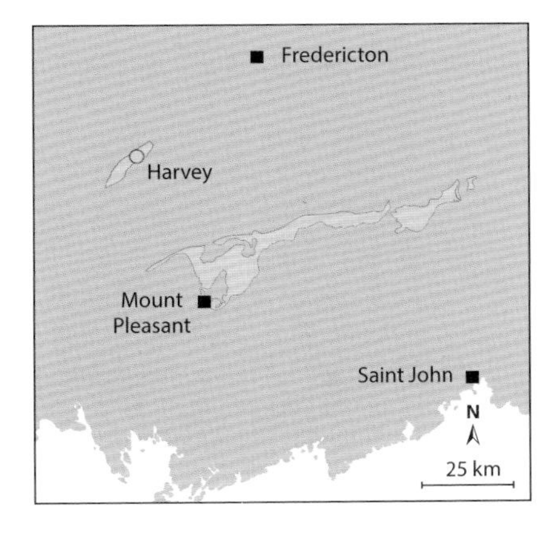

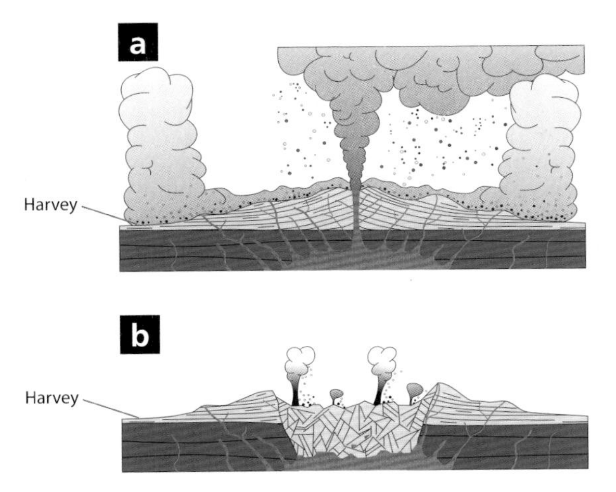

Caldera formation. These two cross-sections through a volcano show (**a**) how a large, explosive eruption can empty the magma chamber beneath a volcano, (**b**) causing the volcanic mountain to collapse and leaving a craterlike caldera in its place.

The grey granite used to build McAdam's historic railway station was locally quarried from boulders.

Boulder Legacy

Glacially Transported Granite on Display

A railway station, a campground, and a rare boulder. What could they have in common? All are in McAdam. All display granite likely derived from the nearby Pokiok batholith that extends northward from the village. All have a connection to the most recent ice age about 20,000 years ago.

When glacial ice last moved across this landscape, it travelled from northwest to southeast, plucking up any loose rocks and carrying them along. Later, as the ice melted, the transported rocks settled to the ground. Because the ice had travelled across a large expanse of granite to get here (see FYI), most of the area's glacial boulders are granite. For example, the village campground is situated in a glacial-era boulder field.

Historically the boulders have served as a practical source of building stone—and plenty of it, as seen in the railway station's vast facade. A 1914 report on quarries in the Maritimes noted of McAdam, "A considerable amount of building stone is produced from the boulders scattered over the country.... Many of the boulders are of large size and yield solid pieces of stone 20 feet [6 metres] in length."

Getting There

Driving Directions

McAdam is on Route 4 about 19 kilometres west of the intersection with Route 3, and about 9 kilometres east of the US border. **Railway station (1)**: Route 4 makes a sharp bend on the north side of town. About 1 kilometre south of the bend, the railway station is on the east side of the road. **Campground (2):** About 500 metres north of the railway station on Route 4 (Saunders Road), turn (N45.59502 W67.32686) west onto Lake Avenue; follow it for about 1 kilometre. Near the lake, the pavement curves right. Stay right and follow the unpaved road into the campground. Watch for signposts.

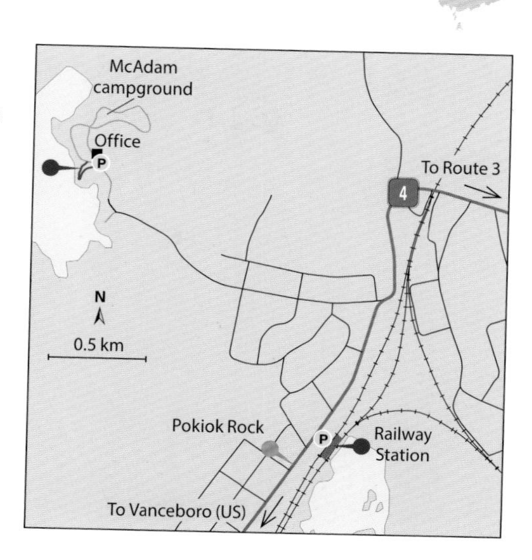

Where to Park

Parking Locations: Railway station (1): N45.59056 W67.32875
Campground (2): N45.59938 W67.33963

(**1**) Park along the side of the wide access lane in front of the station or in an open area nearby. (**2**) Park beside the camp office.

Walking Directions

(**1**) At the station, walk around the outside to view the stone facade. (**2**) At the campground, the Sunset Walking Trail starts near the camp office. From the office, follow the park road that forks left toward the lake. Watch for the signposted trail head on the left. The trail runs roughly parallel to the lakeshore, leads past some benches, then curves left and heads back toward the office.

Notes

For more information about the railway station and the village campground, visit mcadamstation.ca and mcadamnb.com/campground.

1:50,000 Map

McAdam 021G11

Provincial Scenic Route

River Valley Scenic Drive

On the Outcrop

(**a**) The facade of McAdam railway station is made of light grey granite with contrasting red granite accents. (**b**) A trail at the village campground wends through a boulder field.

Outcrop Location: N45.59017 W67.32883

The above waypoint is for the railway station. For the village campground, see Getting There (location 2).

The facade of the railway station makes it easy to get a close view of an undeformed, coarse-grained granite. The stone blocks contain white feldspar, including some well-formed, rectangular crystals 2 centimetres or more in length. Quartz appears in irregular shapes of light grey. Flakes of black biotite are evenly distributed throughout the rock. Some rusty spots mark the site of iron-rich pyrite crystals, and some stone blocks contain small inclusions of darker grey rock.

This same light grey granite can be found at the McAdam campground. Its Sunset Walking Trail (about 250 metres long) and the vehicle lanes that loop around to the campsites (about 1 kilometre) all thread their way through a field of glacially transported granite boulders, some the size of a garden shed. While the boulders at the campground are not as large as those quarried for the railway station, the site provides a sense of the quantity of glacially transported boulders in the area.

Rock Unit

Pokiok Plutonic Suite

Bedrock Map

MP 2005-34

FYI

- The Pokiok batholith underlies an area of about 1,500 square kilometres in southwestern New Brunswick (shown in pink in the map at right). It is a complex of mainly granite intrusions formed between 415 and 402 million years ago. Glacial ice advancing northwest to southeast over the region crossed more than 40 kilometres of nothing but granite before reaching the area where McAdam is now.

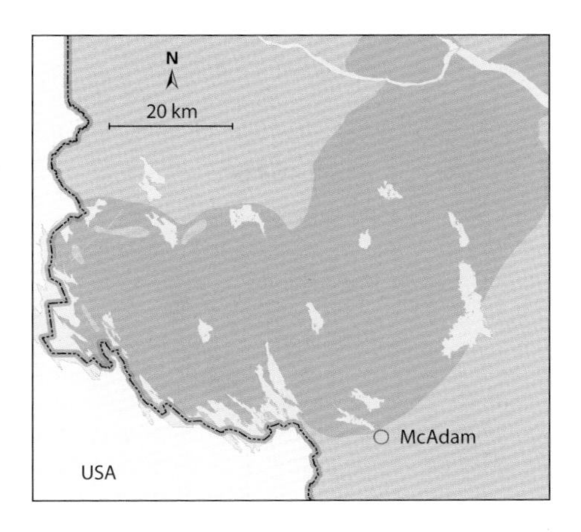

- Construction of the McAdam railway station was commissioned in the late 1800s by W. Cornelius Van Horne, then head of the Canadian Pacific Railway. The journey between his summer home on Ministers Island (site 3) and his principal household in Montreal passed through McAdam.

Also Nearby

On the east side of Saunders Street between the railway station and Pine Street (N45.58948 W67.33033) is a large boulder locally known as Glacier Rock or Pokiok Rock. It's an example of orbicular granite, a rare phenomenon. In the boulder, dark fragments of metamorphosed sedimentary rock are surrounded by pale rims of fine-grained granite 1 to 3 centimetres thick. Holding this outlandish collection together is a matrix of coarse-grained granite. Geologists are still not sure exactly how this kind of orbicular granite forms.

Orbicular granite.

The boulder was discovered along a remote trail north of McAdam. Glacial action must have transported it to that site, but its original source is unknown. No outcrops of similar orbicular granite have yet been found in New Brunswick or neighbouring Maine.

Exploring Further

New Brunswick Department of Energy and Resource Development. "Industrial Minerals Summary Data: McAdam Granite." https://www2.gnb.ca (search "Industrial Minerals Database" then click on the link, and in the search form enter Reference# 94).

Along the Maliseet Trail near Woodstock, 20-metre-high Hays Falls cascades down a cliff face of metamorphosed sedimentary rock.

Calling Card

Cambrian Sediments of the Miramichi Terrane

Come along down the Maliseet Trail near Woodstock, steeped in both human and geological history. The footpath is a vestige of a much longer land and water route developed in pre-colonial times by Indigenous communities of this region. The original trail connected the Saint John River with navigable waters along its tributary, the Eel River. A series of lakes and portages connected the Eel River to Baskahegan Stream and others in the Penobscot watershed of what is now the state of Maine.

Like the Maliseet Trail, the rocks here also reach from the Saint John River into the Penobscot River watershed of Maine. Known as the Baskahegan Lake Formation, these rocks formed during the Cambrian period, while Ganderia was still attached to the margin of Gondwana, on the shore of a widening Iapetus Ocean.

Carried by unnamed rivers reaching into Gondwana's ancient interior, volumes of sediment piled up on the continental shelf and periodically avalanched into deeper water, forming thick sandy layers. Later events heated, pressed, and deformed the layers into the resistant rock seen at Hays Falls.

Getting There

Driving Directions

On Route 2 about 12 kilometres south of Woodstock, take Exit 200 and follow Dugan Road east to its intersection (N46.04379 W67.55564) with Route 165. Turn right (south) and follow Route 165 about 2 kilometres, watching for signs for the hiking trail. At the trail head is a long pull-off on the west side of the road.

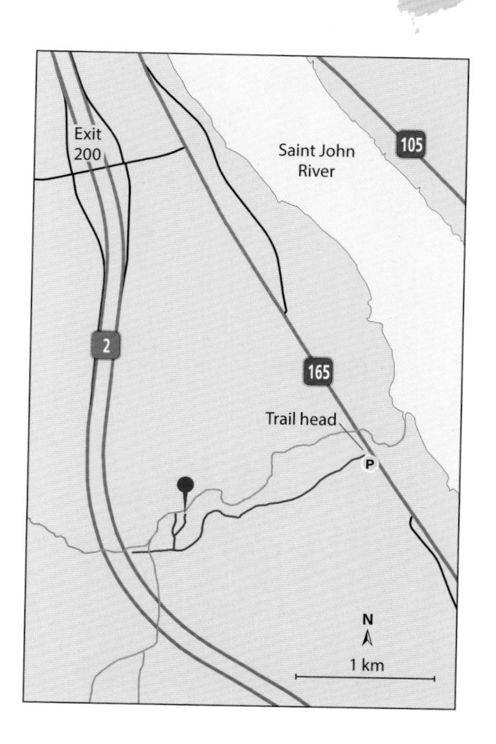

Where to Park

Parking Location: N46.02928 W67.54305

The long pull-off at the trail head has plenty of room for parking.

Walking Directions

Starting at the trail head by the parking location, follow the Maliseet Trail westward for about 1.2 kilometres. At the first fork branch right (north) toward Hays Falls. (Straight ahead the trail continues toward Route 2 and peters out.) You may find that this and the next fork are marked by coloured tape around a nearby tree. Walk about 80 metres to the second fork and bear right for a steep trail to the base of the falls, about another 150 metres.

Notes

The trail is well worn and clearly blazed, but most of it is not improved or maintained. Some stretches are rough with tree roots and rocks, and some are sodden and muddy when the weather has been wet.

1:50,000 Map

Woodstock 021J04

Provincial Scenic Route

River Valley Scenic Drive

On the Outcrop

Thick, tilted beds of quartzite loom over the moss- and fern-clad clearing at the base of Hays Falls.

Outcrop Location: N46.02627 W67.55515

Most of the rocks exposed near the base of the falls are thick-bedded quartzite and metamorphosed greywacke (formed from impure, rough sand containing bits of feldspar and clay rather than pure quartz sand). The grain size of the sediment from which they originally formed varies; some were originally sand and others were silt.

The rocks are very hard because metamorphism has recrystallized the sand and silt grains of the earlier sedimentary rock into firmly interlocking crystals. The jagged edges of the outcrops are a testament to their hard, resistant character.

In several of the outcrops you can see well-defined layers that are tilted in a consistent orientation by large-scale folds. However, in some places, for example on the rock face on the right-hand side at the base of the falls, folding and faulting has left some exposures with a jumble of orientations (marked in the photo at right).

Deformed metasiltstone.

Rock Unit

Baskahegan Lake Formation, Miramichi Group

Bedrock Map

MP 2006-5

FYI

- Microscopic grains of the mineral zircon are very resistant and stable, so they are commonly found in sediment, where they are known as detrital zircon. Because they contain traces of uranium, they can be used for radiometric dating.

Researchers extracted zircon grains from the Baskahegan Lake Formation, sampled along Route 2 near Hays Falls. The zircon ranged in age from 525 million years to more than 2,500 million years. The pattern of ages is a good match for the ancient rocks of Amazonia (now part of South America, but part of Gondwana during the Cambrian and Ordovician periods).

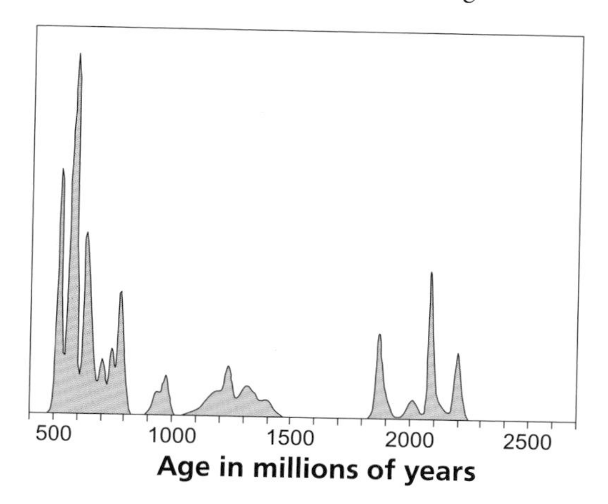

Detrital zircon ages, Baskahegan Lake Formation.

- Throughout the Cambrian period and early in the Ordovician period, sand and silt accumulated along the margin of Ganderia that was facing Gondwana and on the margin of a back-arc basin that formed within Ganderia. The rocks formed from those sediments are now folded and metamorphosed, occurring in a broad band across New Brunswick. They are mainly found in the Miramichi terrane but also in the St. Croix terrane.

The youngest part of this deepwater sequence commonly takes the form of dark, sometimes black, shale, examples of which have been found within 10 kilometres of Hays Falls. Those who study Appalachian geology consider this combination (thick Cambrian greywacke followed by a cap of early Ordovician black shale) to be the "geological calling card" of Ganderia. A similar sequence is found along the Nepisiguit River south of Bathurst (see site 38, Middle Landing).

HF, Hays Falls; ML, Middle Landing.

The Malabeam Tourist Information Centre (right) is perched beside the rocky gorge at Grand Falls.

Folds and Refolds

Ordovician-Silurian Rocks of the Matapédia Basin

Grand Falls and its river gorge are breathtaking. The sheer scale of the scenery has few equals in Atlantic Canada, drawing you from one dramatic view to the next. From clifftop trails, stairways, ziplines, and river tours, you can view the water and rocky precipices from every angle. So the part where you are happily awed by the rocks comes first.

Then comes the part where the rocks make you curious. The gorge exposes thousands of thin, grey layers, stacked on edge like an endless deck of cards. In some places the layers swirl and curve in complex forms. And swarms of bright, white veins gash the cliffs. How did all this come about?

It started late in the Ordovician period, when a deep marine depression—the Matapédia basin—formed near the margin of Laurentia. Underwater currents brought small but persistent amounts of mud and silt into the basin, layer by layer, and the sediment accumulated implacably for millions of years. That is, until the Salinic and Acadian orogenies came along with a new idea.

Getting There

Driving Directions

From Route 2, take Exit 75 and follow Route 108 into town. As you near the river, watch for the Malabeam Tourist Information Centre on the right, and turn into the parking lot.

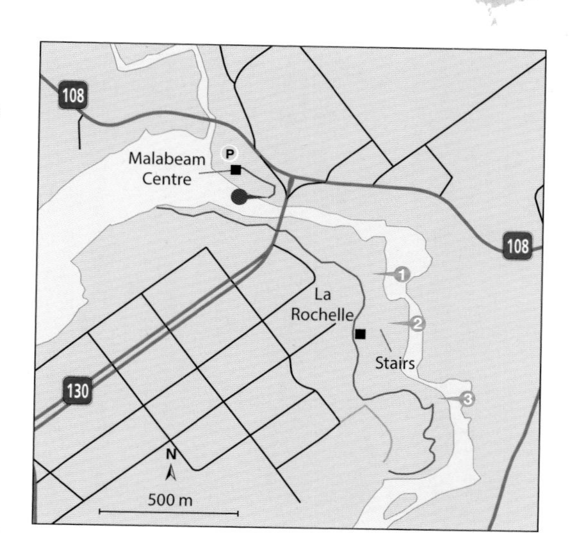

Where to Park

Parking Location: N47.05204 W67.73825

The Malabeam Centre has ample parking in its paved lot. For access to downstream trails, you may prefer to park at La Rochelle Tourist Centre (N47.04783 W67.73381).

Walking Directions

In the parking lot, as you face the Malabeam Centre, follow the sidewalk to the left, continuing past the end of the parking lot onto a boardwalk and then a metal walkway. These lead to a lookout at the edge of the river gorge (N47.05098 W67.73678). To access additional trails and lookouts across the river, follow sidewalks to and across the nearby Ron Turcotte Bridge. A brick walkway curves from the upstream side of the bridge down to the trail.

Upstream, the trail leads to lookouts over the dam. Downstream it follows the river for about 1 kilometre. Other locations with expansive views of the rocky gorge and riverbed include (**1**) a signposted lookout about 250 metres southeast of the bridge (N47.04920 W67.73347), (**2**) a remarkable stairway system accessed via La Rochelle Tourist Centre, and (**3**) the Falls Brook lookout (N47.04634 W67.73126) near the town campground.

Notes

Low-water conditions are best for viewing rock features. The trail route includes gravel, boardwalk, and natural surfaces. For more information about Grand Falls, including a map of town amenities, visit www.grandfallsnb.com.

1:50,000 Map

Grand Falls 021O04

Provincial Scenic Route

River Valley Scenic Drive

On the Outcrop

At the lookout by the Malabeam Centre are large expanses of nearly upright layers. In contrast, complex distortions can be seen in the rocks at lookouts farther downstream.

Outcrop Location: N47.05098 W67.73678

From the lookout near the Malabeam Centre (waypoint above), you can clearly see the thin layers of calcite-rich mudstone that characterize this part of the Matapédia basin. The layers are now steeply tilted, exposing their edges. The proportion of calcite in the mudstone varies from one layer to the next, and differences in their weathering rates have left the ridged patterns in the rock surface.

Originally near horizontal when deposited in the basin, the layers are now nearly vertical and aligned roughly northeast-southwest due to intense folding (see FYI). The orientation of the layers is consistent over a large area, as you can see if you follow the trail downstream. These features are typical of deformation caused by the Acadian orogeny.

At downstream lookouts 1 or 2 (see Getting There), in low-water conditions you can see that some rock layers exposed in the riverbed are distorted into complex shapes. This suggests multiple episodes of deformation. Although the Acadian orogeny's strong imprint of vertical folding dominates the falls and gorge, in a few places like these the earlier effects of the Salinic orogeny survive as refolded folds.

Rock Unit

White Head Formation,
Matapédia Group

Bedrock Map

NR-1 (2008)

FYI

- On several cliff faces and outcrops, numerous bright white veins of calcite are a conspicuous feature. These formed during folding as calcium-carbonate-rich fluids escaped from the rocks.
- The calcite in these rocks mainly takes the form of fine particles dispersed throughout the rock. Much of this material was eroded from the continental margin of Laurentia, where an extensive carbonate shelf had formed in warm tropical waters of the Iapetus Ocean during the Cambrian period.
- Because its tectonic setting changed over time, the Matapédia basin is known as a composite basin. Its sedimentary layers accumulated over a very long period, from about 450 to about 405 million years. They now cover a large area in northwestern New Brunswick (including site 31, Edmundston) as well as neighbouring Quebec and Maine.

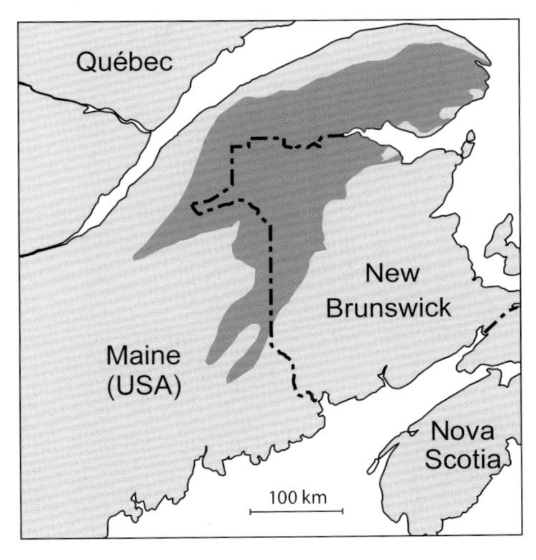

Matapédia basin.

- The Acadian orogeny is known for having caused upright folds to form over a very wide area. Caused by the collision of Avalonia with Ganderia, the effects of the orogeny moved gradually inland from the point of collision. Upright folds can leave once-horizontal layers nearly vertical (see diagram).

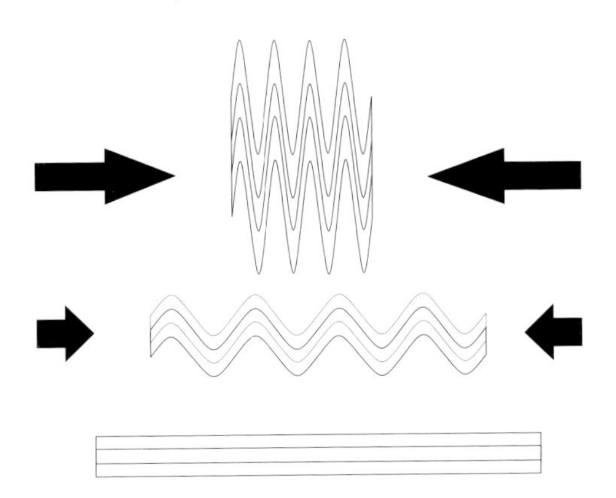

Edmundston's reconstructed Blockhouse at Little Falls rises above the surrounding landscape, thanks to the rocky knoll on which it is built.

Fold Features

Devonian Shale Transformed into Slate

Nearly 150 years after Edmundston's blockhouse burned down, local history enthusiasts found archival plans, spruced up the original foundation, and put great care into building this accurate replica. The structure is unusual in the angular offset of its third storey, which allowed its occupants to monitor—and defend against—intruders approaching along either nearby river.

The bedrock on which the blockhouse is built is also a study in angular complexity. Originally formed from layer upon layer of fine silt and clay, the rock was later folded and transformed into slate. In slate, tiny, flaky clay minerals have recrystallized in alignment, resulting in surfaces along which the rock can be easily split, or cleaved. This feature, cleavage, is aligned vertically at the blockhouse, while the original sedimentary layering is roughly horizontal (see FYI).

The foundation and first storey of the blockhouse are made of similar slate. In them, you can see the original sedimentary layers preserved as colour contrasts (from light to dark grey or brown). These colours form intricate patterns in some of the building stones, depending on how the slate broke across the sedimentary layers.

Getting There

Driving Directions

If travelling on Route 2, in Edmundston take Exit 13 or 21 to Route 144 and drive toward the city centre. Just east of the Matawaska River, watch for the intersection with Avenue St-Jean (N47.36293 W68.32198).

If travelling from the US, follow Route 1 in Madawaska to Bridge Avenue and cross into Canada. Turn right onto Rue St-Francois, then right again onto Route 144. Continue across a smaller bridge and watch for Avenue St-Jean.

Turn north onto Avenue St-Jean, and about 50 metres from the intersection, turn left (west) into the parking area for the Blockhouse at Little Falls.

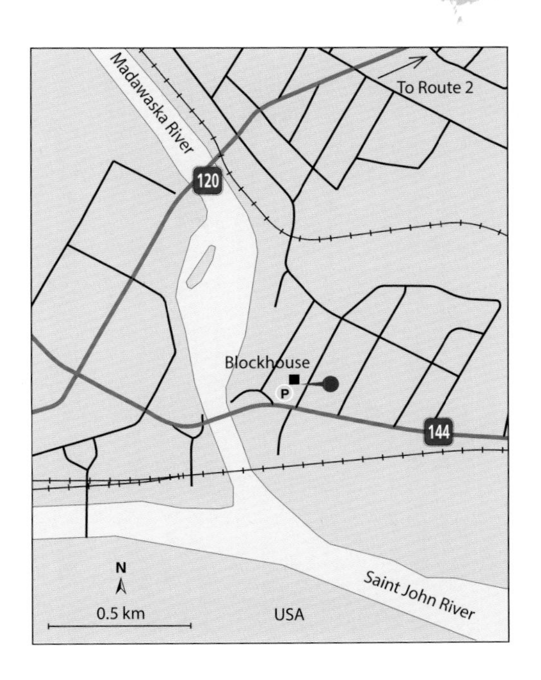

Where to Park

Parking Location: N47.36341 W68.32232

The parking lot is on the south side of the blockhouse.

Walking Directions

From the parking location walk toward the blockhouse to view the rocky prominence on which the building sits. From the sidewalk along Avenue St-Jean you can view a small rock face and gain easy access to the rock pavement near the building foundation.

Notes

The Blockhouse at Little Falls is a registered historic site. There is no cost to explore the grounds. For more information, visit www.historicplaces.ca/en/rep-reg/place-lieu.aspx?id=2156.

1:50,000 Map

Edmundston 021N08

Provincial Scenic Route

River Valley Scenic Drive

On the Outcrop

In the outcrop below the blockhouse, the original sedimentary layering is almost at right angles to the slate's nearly vertical cleavage (see arrows).

Outcrop Location: N47.36350 W68.32199

Features of this rock are best seen along vertical edges, which can be found around the blockhouse and along the sidewalk on Avenue St-Jean. The original sedimentary layering is slightly tilted but nearly horizontal. On some surfaces it can be recognized from differences in how the layers have weathered (see photo above); on fresher surfaces you may notice colour variations in shades of brown or grey.

Vertical cleavage.

Unlike the sedimentary layers, the cleavage caused by deformation of the rocks is nearly vertical and is aligned approximately northeast-southwest. This feature is best seen along northeast- or southwest-facing parts of the outcrop, where it appears as thin vertical ridges along the rock surface. Outcrops along the sidewalk clearly display this feature (photo at right).

Rock Unit	**Bedrock Map**
Témiscouata Formation, Fortin Group	MP 75-166 (no colour)

FYI

- Geologists describe the orientation of any type of rock layering, including cleavage or sedimentary layers, both in terms of its dip (horizontal, vertical, or somewhere in between) and its strike (a compass direction). In outcrops around the blockhouse, the strike of the cleavage is about 40 degrees, roughly northeast; its dip is 90 degrees or slightly less, roughly vertical.

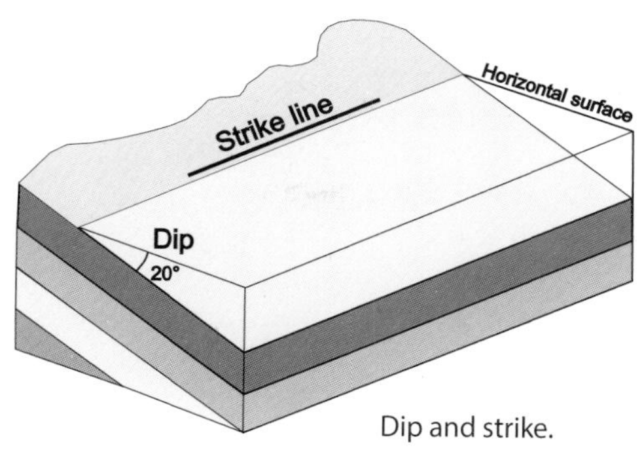

Dip and strike.

- The angle between folded sedimentary layering and cleavage may vary from one part of a fold to another, depending on the shape of the fold and the orientation of the tectonic forces squeezing the rocks. In upright folds like those caused by the Acadian orogeny, the two features are likely to be nearly perpendicular—like they are at this site—in the "noses" (top and bottom curves) of folds.

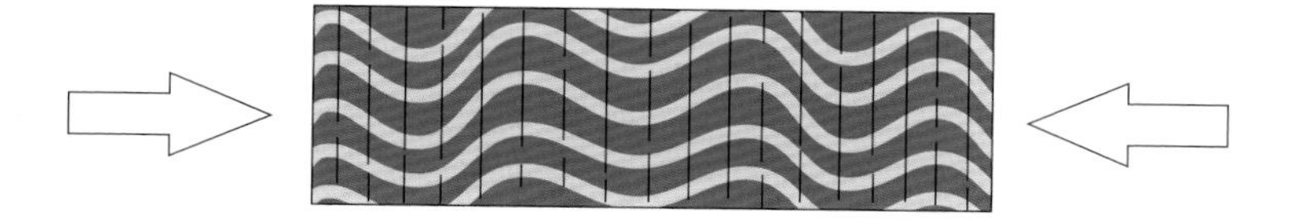

- The rocks in Edmundston originally formed as part of the later history of the Matapédia basin, a low-lying area that persisted for tens of millions of years in various forms. The rocks at site 30, Grand Falls, formed early in the history of this same basin.

- The fine grain size of the original sediment, combined with the uniform cleavage imposed during the Acadian orogeny, makes this rock formation well suited for the production of high-quality building slate. A quarry operation is located nearby in Saint-Marc-du-Lac-Long, Quebec. It has provided slate for St. Michael's Cathedral in Toronto, McGill University, and La Sorbonne in Paris, among others.

Exploring Further

More information about the nearby slate quarry in this same rock formation in Quebec:

Glendyne, Inc., https://www.glendyne.com/en/gisement.htm (slate qualities) or https://www.glendyne.com/en/realisations.htm (list of international customers).

Sagamook Mountain rises beside Nictau Lake in Mount Carleton Provincial Park.

Underwater Eruptions

Signs of Devonian Unrest

Mount Carleton Provincial Park provides public access to a unique wild area of New Brunswick. Expansive lakes, thriving Acadian woodland, the Maritimes' highest mountains, and its status as a dark sky preserve bring visitors close to the natural world. Rock exposures are plentiful on the mountain peaks, but you don't have to be a seasoned hiker to experience key elements of the park's geological past.

Scenic, rocky Williams Falls is just a short hike from the park road, and along the way are additional outcrops beside the road. As you travel to the falls, you can examine three of the rock types in the park, all formed in a tectonically restless setting early in the Devonian period: granite (outcrop 1), mudstone (outcrop 2), and felsic tuff (outcrop 3).

At the time these rocks formed, Ganderia had completed its collision with Laurentia; Avalonia had joined them, too. But peace and quiet were not yet in the cards as the Devonian period began. Complex plate interactions caused a basin to form here, and in the basin, underwater volcanoes erupted mainly felsic lava and ash. That's uncommon—most underwater volcanic activity is mafic.

Getting There

Driving Directions

Along Route 180 about 30 kilometres east of St. Quentin, watch for the intersection (N47.48939 W67.00640) with Route 385 and turn south. Follow Route 385 for about 9 kilometres and watch for the park entrance on the left (N47.43004 W66.94803). Register at the park kiosk, then continue along the park road. In about 1.2 kilometres fork left to cross the river. Follow the park road on the north side of Lake Nictau, travelling about 6.4 kilometres from the park office to the trail head for Williams Falls. Visit two roadside sites along the way (4.6 and 6 kilometres from the park office).

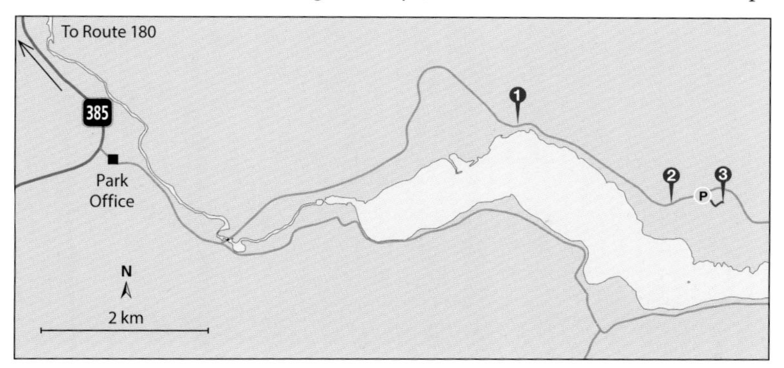

Where to Park

Parking Location: Williams Falls (**3**): N47.42659 W66.88368

(**1, 2**) The first two outcrops are beside the park road. Pull off to avoid blocking traffic. (**3**) At the trail head for Williams Falls is a parking area with room for several cars.

Walking Directions

(**3**) From the parking location at the trail head for Williams Falls, follow the clearly defined and well-maintained trail, which winds through the forest toward the waterfall. Cross the bridge above the falls and follow the trail around to view the front of the cascade.

Notes

This site lies within the boundaries of Mt. Carleton Provincial Park and requires a valid park pass. For more information about the park, visit www.tourismnewbrunswick.ca (search for "Mount Carleton").

1:50,000 Map

Nepisiguit Lakes 021O07

Provincial Scenic Route

Appalachian Range Route

On the Outcrop (1)

A fractured outcrop along the park road provides a look at the Mount Bailey granite.

Outcrop Location: N47.43178 W66.90342

About 4.6 kilometres from the park office, a pull-off on the south side of the road provides a view across the lake to Sagamook Mountain. Across the road is a dark pink or brick-coloured outcrop. It has a rubbly, blocky appearance with sharp broken edges.

Very fine-grained granite.

This rock is definitely igneous, but was it erupted or intruded? Geologists have examined its minerals under a microscope to find out. It contains mainly quartz and feldspar, and based on the mineral textures it is considered to be a fine-grained granite. It is so fine grained that it fractures much like flint.

The outcrop is part of a narrow band of similar granite trending roughly north to south (pale pink in the map on p. 177, FYI). Likely it represents a "feeder" vent near the surface—magma that was on its way to an eruption, but never made it. Sometimes such "almost erupted" rocks are referred to as subvolcanic rocks.

Rock Unit	**Bedrock Map**
Mount Bailey Granite	MP 2012-60

On the Outcrop (2)

In a roadside outcrop near the trail head for Williams Falls, fine-grained mudstone shows the effects of deformation.

Outcrop Location: N47.42604 W66.88691

About 6 kilometres from the park office, just past the access road for a park campground on the south side of the road, watch the north side for a low, dark outcrop behind a stand of young birch.

These rocks originated as fine-grained marine sedimentary rock, mudstone. The sediment accumulated during quiet periods between volcanic episodes in the basin. In some other locations in the park, similar sedimentary rock contains volcanic fragments, showing that the sedimentary and volcanic processes were going on together.

Vertical cleavage in shale.

Later, when the rocks were deformed, areas of hard igneous rock nearby (for example, those in outcrops 1 and 3) were less affected. Extra-severe deformation of the mudstone made up the difference. This left the mudstone with a near-vertical fabric known as cleavage. See site 31, Edmundston, FYI, for more information about this rock property.

Rock Unit

Wapske Formation, Tobique Group

Bedrock Map

MP 2012-60

On the Outcrop (3)

Williams Falls tumbles over a ledge of resistant volcanic rock.

Outcrop Location: N47.42607 W66.88146

The rocks over which Williams Brook tumbles here are part of a thick layer of felsic volcanic ash and debris. This silica-rich rock is hard; its resistance to erosion has led to the formation of the falls. Many of the outcrops by the falls are obscured by lichen or moss, revealing few details. Look for patches where the rock has broken or the moss has been pulled away (but don't damage the site to make new exposures!). Foot-worn outcrops in the trail itself are also revealing.

Geologists who have studied this area describe the rocks as waterlain (or subaqueous) tuff, meaning the material in them accumulated underwater. On some surfaces where details are clear, crystal fragments (about 1 millimetre across) and larger, irregular fragments are visible in the fine matrix of ash.

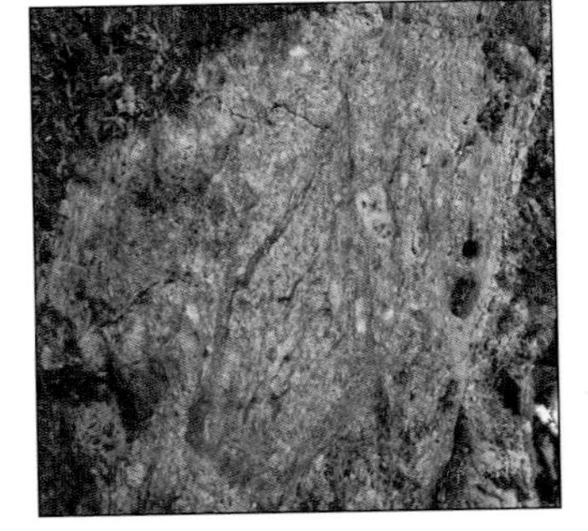

Waterlain tuff.

Geologists call these hyaloclastic fragments. They form in an underwater eruption when water touches the exposed surface of hot magma. A hard, glassy rim forms instantly, then promptly shatters.

Rock Unit

Wapske Formation, Tobique Group

Bedrock Map

MP 2012-60

FYI

- The marine basin in which these volcanic and sedimentary rocks formed was located on continental crust, not on ocean crust. A sequence of collisions during the Silurian period left the region unstable. Numerous large faults cut the region into blocks that jostled and elbowed each other, warping the crust.
- Mount Carleton is the highest peak in the Maritimes at 820 metres. At the summit are extensive exposures of rhyolite closely related to the felsic volcanic rocks at Williams Falls.

Summit of Mount Carleton.

Also Nearby

- Nearly all of the park's major peaks (green triangles in the map below), and all points above 600 metres elevation (inside the green line) occur within areas of hard, fine-grained granite (pale pink) or rhyolite (pink), which are resistant to weathering and erosion. Areas of sedimentary rock (darker brown) are flatter and lower. A hilly area of mixed sedimentary and volcanic layers (light brown) lies between them in the north (rivers and lakes in pale blue).

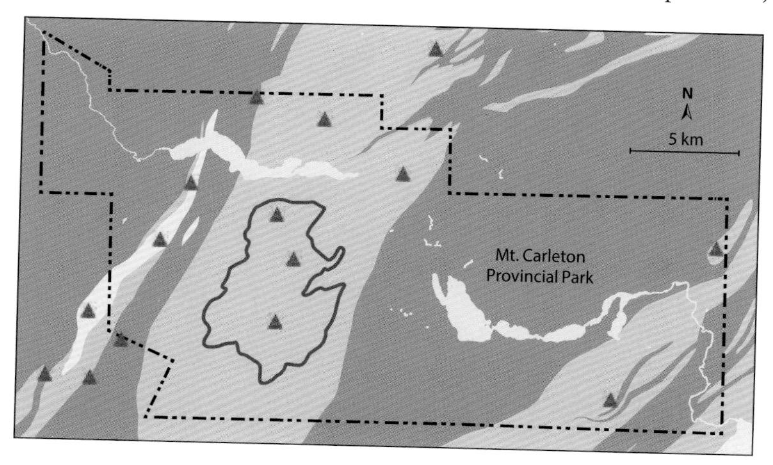

Sugarloaf Provincial Park in Campbellton is named for Sugarloaf Mountain, seen here looking northeast from a park road.

Not Quite Erupted

Conduit beneath a Devonian Volcano

Sugarloaf Mountain rises abruptly from the surrounding landscape in Campbellton. It provides a well-known landmark for the community, a wonderful workout for hikers, an amazing view across the Restigouche River into Quebec, and a great geological story.

A popular misconception is that Sugarloaf is an extinct volcano. That's not quite right in the geological sense, in which "volcano" refers to a mountain built up around a volcanic vent by the accumulation of erupted material. There are no lava flows, volcanic ash, or other products of volcanic eruption on the mountain. Instead, the rocks of Sugarloaf formed part of the underground plumbing system of a volcano that once existed some distance above the current site.

Early in the Devonian period, Laurentia, Ganderia, and Avalonia formed a continuous expanse of continental crust, but the region was still tectonically active and unstable. In some regions, including this one, heat poured upward from the mantle, melting the lower crust to form pools of magma. The magma rose upward and some did erupt. But the rocks of Sugarloaf itself never made it to the surface.

Getting There

Driving Directions

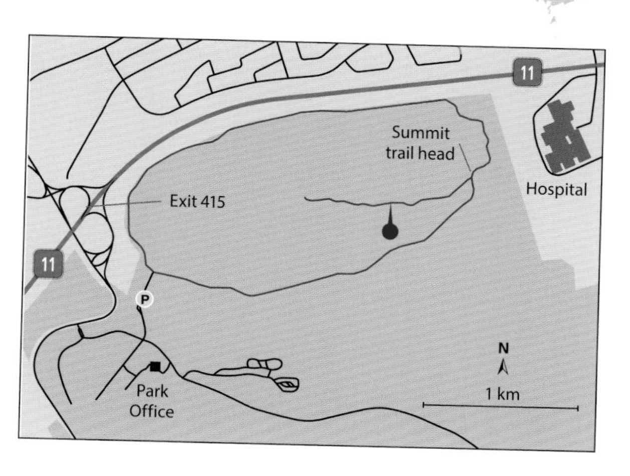

From Route 11 in Campbellton, take Exit 415 and turn left onto Val d'Amour Road, following signs for Sugarloaf Provincial Park to the park entrance (N47.98446 W66.69815). From the entrance, follow Athollville Street around to the left and at the intersection with Campbellton Street, turn left again. Continue about 150 metres to the parking area.

Where to Park

Parking Location: N47.98565 W66.69408

There is plenty of room available for parking between Campbellton Street and the pond.

Walking Directions

From the parking location by the pond, walk toward Sugarloaf along the gravel path. You may see signs for the Terry Fox Trail, which encircles the base of the mountain. At the monument, turn right on the Terry Fox Trail and follow it for about 1.6 kilometres to the trail head for the summit trail. At that point you will be near the grounds of the regional hospital.

Follow the summit trail at least to the outcrop location, about 400 metres from the Terry Fox Trail. Continue to the summit (about 800 metres from the Terry Fox Trail) for more outcrops and a great view of Campbellton, the Restigouche valley, and neighbouring Quebec.

Notes

Sugarloaf Provincial Park is open year-round. For more information, visit http://parcsugarloafpark.ca.

1:50,000 Map

Atholville 021O15

Provincial Scenic Route

Appalachian Range Route

On the Outcrop

Part of the summit trail on Sugarloaf Mountain leads over fractured bedrock.

Outcrop Location: N47.98974 W66.68010

About 400 metres from the Terry Fox Trail, the summit trail crosses an unusual looking outcrop of fractured rock (photo above). The rock is broken into roughly parallel, slightly tilted layers. The natural phenomenon that caused these fractures is known as columnar jointing.

On freshly broken surfaces you can see that the rock has a fine-grained grey matrix containing visible crystals up to about 5 millimetres long. The white rectangular crystals are feldspar; the long, narrow black ones have been altered but were probably amphibole or biotite.

Trachyte porphyry.

On flat rock surfaces, look for a very thin layer of reddish stain. Such stains commonly form on the outside surfaces of igneous columns as mineral-bearing fluids circulate along the fractures. A less common, more subtle feature of some column surfaces is called a plumose fracture. This appears like a shallow, scallop- or plume-shaped pattern on the rock surface, up to 10 centimetres wide (see FYI).

Rock Unit

Sugarloaf Porphyry

Bedrock Map

MP 2013-15

FYI

- The rock on Sugarloaf Mountain is closely related to the rock at Inch Arran Point in Dalhousie (site 34). Both are a type called trachyte porphyry. Trachyte is similar to rhyolite, but with little or no quartz.

  A porphyry is an igneous rock in which a very fine-grained matrix surrounds scattered, individual crystals that are visible to the unaided eye. The larger crystals formed slowly while the rock was still molten. With just liquid surrounding them, they were free to grow in their natural shape. Later, when the magma moved quickly upward into cooler regions near the Earth's surface, the remaining melt crystallized so quickly that the mineral grains are microscopic.

- Magma that pools in large volumes on or near the Earth's surface cools from the outside inward. Because cooling causes shrinkage, in some cases a network of cracks grows as cooling progresses, eventually forming columns. The columns are not always vertical; their orientation depends on the details of how the rock cooled. Sites 8 (Southwest Head) and 25 (Currie Mountain) are other New Brunswick locations where you can see columnar joints.

This rock surface on Sugarloaf Mountain has two features typical of columnar joints. On the right, a thin coating of rusty red stains the surface. On the left, a subtle scallop- or plume-shaped pattern of radiating ridges and grooves marks the location of a plumose fracture.

Exploring Further

University of Oregon. "Columnar Jointing." *Volcano World.* http://volcano.oregonstate.edu/columnar-jointing.

Wilson, Reg. "Sugarloaf Mountain." *Magnificent Rocks.* https://www.youtube.com/watch?v=infDBccTbQY (explanation by the geologist who mapped this area).

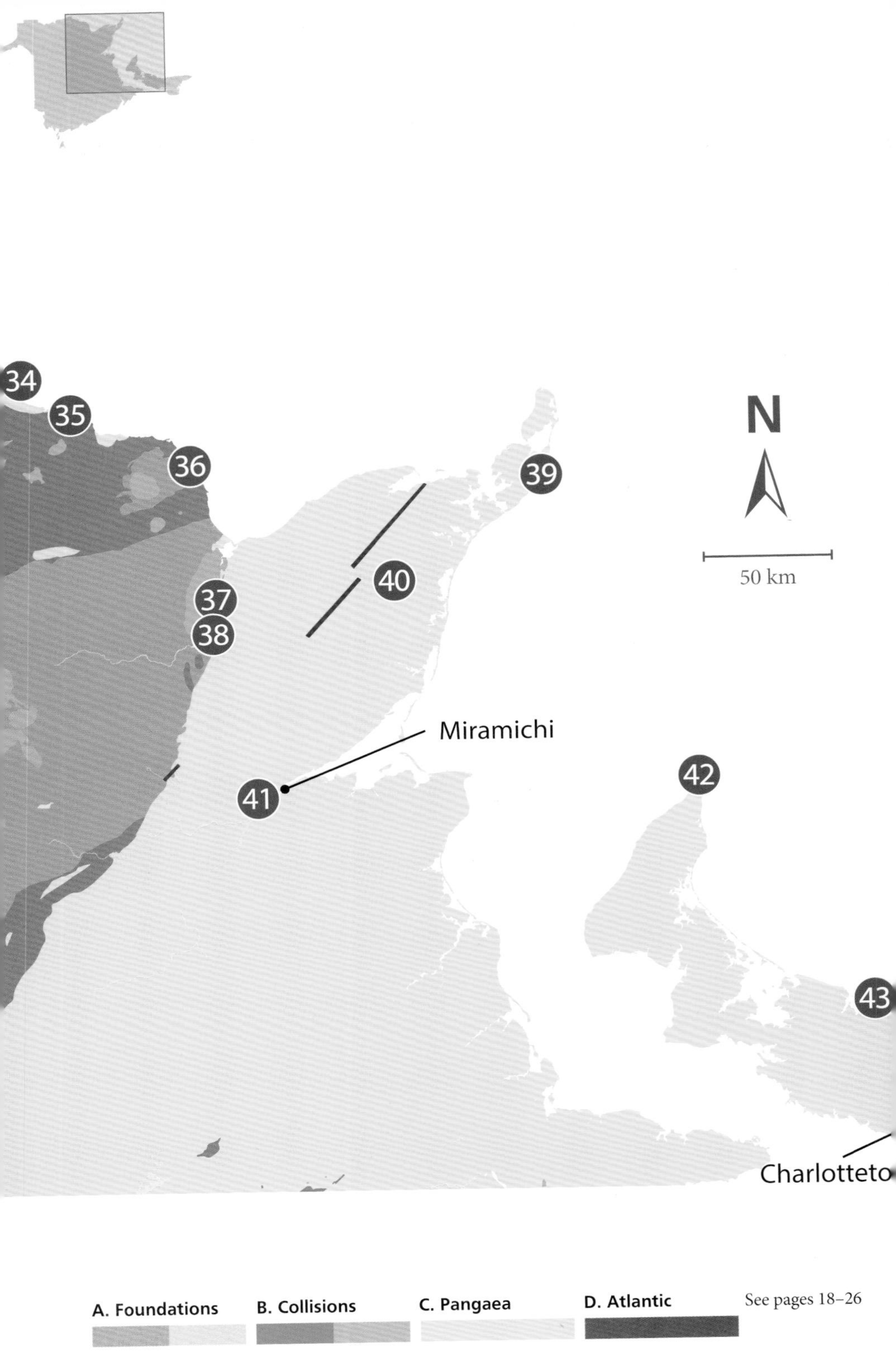

34
35
36
37
38
39
40
41
42
43
N
50 km
Miramichi
Charlotteto
A. Foundations
B. Collisions
C. Pangaea
D. Atlantic
See pages 18–26

NORTHEAST

Tour 4 at a Glance

Tour 4 follows Highway 11 around the eastern shores of New Brunswick, then crosses to Prince Edward Island for a close look at its signature red cliffs. From forces that made mountains to forces that wore them down, the region tells a compelling tale of continental collision and its aftermath. The tour also celebrates the two provinces' rich heritage of stone buildings, made possible by their geological history.

At these sites, you can explore ...

34	**Inch Arran Point**	Igneous rocks of unusual composition
35	**West Point Island**	A volcano-derived sedimentary rock
36	**Atlas Park**	A slice of gabbro from a long-lost ocean floor
37	**Pabineau Falls**	Undeformed Devonian granite
38	**Middle Landing**	Highly deformed Ordovician shale
39	**Cap-Bateau**	A colour change late in the Carboniferous
40	**Grande-Anse to Neguac**	The Acadian Peninsula's sandstone heritage
41	**French Fort Cove**	Carboniferous recycling of sand
42	**North Cape**	Sandstone from a time of climate change
43	**Cavendish Beach**	Prince Edward Island's youngest sandstone
44	**Charlottetown**	Historic buildings of local Permian sandstone

The lighthouse on rocky Inch Arran Point is still in use, guiding traffic into Dalhousie harbour.

Tracking the Trachyte

Igneous Rocks of Unusual Composition

Inch Arran Point was named by Scottish settlers in Dalhousie who came from the Isle of Arran. In Scottish dialect, *inch* refers to a small island or point of land. The lighthouse on the point has been New Brunswick's most northerly beacon since its construction in 1870. It is a fraction of a degree farther north than the lighthouse on Miscou Island at the tip of the Acadian Peninsula.

Views of this picturesque lighthouse against the sky and the broad waters of Chaleur Bay certainly deserve your attention. But as part of your visit, you may also want to cast your eyes downward. The rocks here are closely related to those of Sugarloaf Mountain (site 33) in Campbellton, about 30 kilometres to the west, and share their unusual composition.

These rocks and others of similar age (see FYI) illustrate the tectonic malaise that affected a wide region early in the Devonian period. Ganderia, already wedged against Laurentia, was rear-ended by Avalonia. The effects reverberated throughout what is now the Maritimes. It also seems likely that the Earth's mantle was all astir deep below, providing a flow of heat to the region that caused recurring volcanic activity.

Getting There

Driving Directions

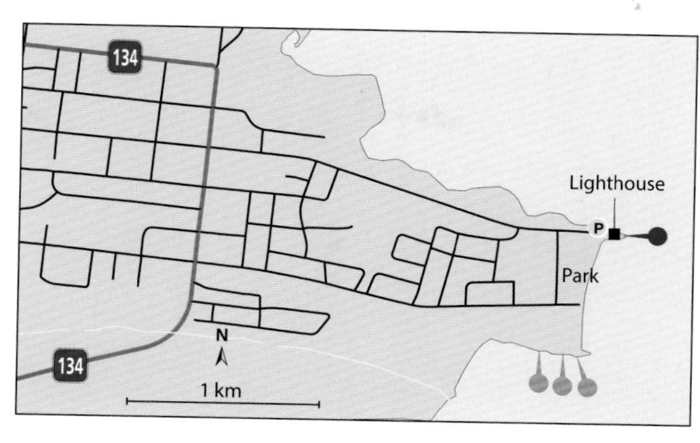

From Route 11, Route 134 in Dalhousie can be accessed from Exit 397 west of town or from Exit 391 (via Route 275) south of town.

From the west, Route 134 enters Dalhousie on Victoria Street. Where it meets George Street, instead of turning left to follow Route 134, continue straight on Victoria Street to its end point at the lighthouse.

From the south, Route 134 intersects Victoria Street just downhill from the large brick buildings and playing fields of the Académie Notre-Dame. Turn right (east) and follow Victoria Street to the lighthouse.

Where to Park

Parking Location: N48.06115 W66.35166

Victoria Street ends in a large paved parking lot by the lighthouse.

Walking Directions

From the parking area, walk across the grass, passing around the left side of the lighthouse. As conditions allow, approach the edge of the grass to the left and in front of the lighthouse.

Notes

The lighthouse is listed on Canada's Register of Historic Places. For more information about its history, visit www.historicplaces.ca and search for listing no. 21094.

Dalhousie's Inch Arran Park, just south of the point, has a sandy beach and amenities for camping, picnics, etc. For more information about the park, visit www.dalhousie.ca (attractions & facilities).

1:50,000 Map

Escuminac 022B01

Provincial Scenic Route

Appalachian Range Route

On the Outcrop

Tilted layers of light brown porphyry border the lush grass in front of Inch Arran lighthouse.

Outcrop Location: N48.06097 W66.35057

As conditions allow, get near the outcrops on the point to view the rock closely. The "polka-dot" appearance of this porphyry is due to small but clearly visible crystals of white plagioclase feldspar in a light brown matrix of microscopic, strongly aligned potassium feldspar with lesser amounts of a dark mineral, pyroxene.

This alignment of crystals in the matrix formed as magma flowed while cooling. It caused the subtle layering in the rocks, which you can see, for example, along the north side of the point. The rock probably represents the remains of a sill (intruded between layers of pre-existing rock near the surface), but it might have been a lava flow.

Trachyte porphyry.

Because the rocks are rather light in colour, with plenty of feldspar and few dark minerals, the outcrop resembles a typical felsic porphyry. But this is trachyte, an unusual rock type (see FYI).

Rock Unit

Val d'Amour Formation, Dalhousie Group

Bedrock Map

MP 2012-53

FYI

- Igneous rocks that are rich in feldspar typically also contain quartz. But trachyte contains little or no quartz, just lots of feldspar. Why?

  Feldspar is a complex family of minerals that contain mainly silica, aluminum, sodium, and potassium. Quartz is just silica. As magma cools, minerals always crystallize in the same order, feldspar before quartz.

  In the rocks at Inch Arran Point, the magma had proportionally so much aluminum, sodium, and potassium that when feldspar crystallized, it used up almost all the silica. By the time the magma was cool enough for quartz to form, there was little or no silica left.

- During the Devonian period, igneous activity was widespread across what had been Ganderia. Beginning with localized activity late in the Silurian period (darkest brown in the map at right), intrusions and eruptions greatly increased early in the Devonian period (brown), then waned later in the Devonian (light brown and tan).

Also Nearby

From nearby Inch Arran Park, at low tide it is possible to walk from the south end of the beach around the point to view the rocks along the shoreline. Separated from the rocks at the lighthouse by a fault, they are noticeably different.

The most conspicuous feature is a small sea arch of basalt. Along the shore near the arch, rock outcrops represent a complex mixture of mafic volcanic rocks and sedimentary rocks. In some outcrops the two occur together, forming peperite (see site 13, Taylors Island).

Basalt arch.

Exploring Further

Wilson, Reginald, and others. "Day 3: Campbellton to Bathurst. Stop 43." *New Brunswick Appalachian Transect: Bedrock and Quaternary Geology of the Mount Carleton-Restigouche River Area.* Fredericton: New Brunswick Department of Natural Resources, 2005, pp. 65–68. https://www2.gnb.ca (search "Appalachian transect" and click on the resulting link). This field guide written by and for geologists includes a detailed description of the rocks near the sea arch.

Don't underestimate this modest-looking conglomerate outcrop by the wharf at West Point Island; it contains interesting volcanic fragments.

Dual Identity

A Volcano-Derived Sedimentary Rock

Athletes, books, cats, dogs—the list of things humans enthuse about is long and varied; it could take you through the alphabet. Surely your own list will include conglomerate under C. It's a rock type whose appearance and history can both be elaborate, a wealth of information in a form that is also enjoyable to look at.

The conglomerate at this site suggests a lively scene. Most of the clasts (rock fragments) are examples of rhyolite in various forms (which match known volcanic rocks nearby). Many clasts are large, apparently not having travelled far. Sedimentary features of the conglomerate suggest rapid deposition, perhaps even as landslides, which may have been triggered by ongoing eruptions.

Formed at the end of the Silurian period, the rocks by the wharf are part of a complex time in the geologic history of the region. The Salinic orogeny was waning; the Acadian orogeny was nascent. It has proven difficult to sort out the effects of each, but certainly one of the outcomes was a flurry of late Silurian volcanic activity in what is now northern New Brunswick.

Getting There

Driving Directions

Route 134 near New Mills can be accessed from Route 11 at Exits 375 or 357. From Route 134 about 1.5 kilometres northwest of the Benjamin River, watch for the intersection with West Point Road (N47.96942 W66.18823) and turn northeast. At the railway crossing, use caution and watch for train traffic. The crossing has lights but no gate. Follow West Point Road as it curves left, then right, and crosses a small causeway onto West Point Island. Continue to the end of the road at the wharf.

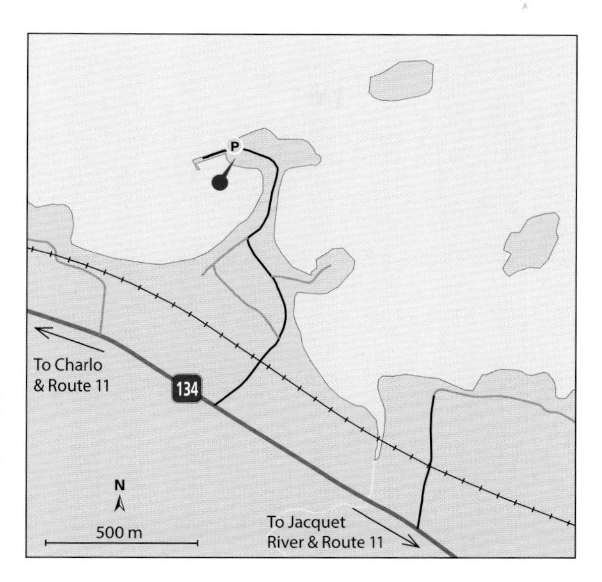

Where to Park

Parking Location: N47.97478 W66.18752

Park near the public wharf, being careful not to block access or encroach on adjacent private property.

Walking Directions

From West Point Road facing the wharf, turn left (south) to see a small rocky outcrop about 25 metres away. The east side facing away from the wharf offers the best view of the rock features.

Notes

The area in front of the rock face is rough with rock debris.

1:50,000 Map

Charlo 021O16

Provincial Scenic Route

Acadian Coastal Drive

On the Outcrop

(**a**) The east side of the outcrop is well exposed and features (**b**, **c**) volcanic debris (mainly rhyolite clasts) in a wide range of sizes.

Outcrop Location: N47.97457 W66.18758

In this outcrop, the rock face on the east side (photo **a**) cuts through several layers, which tilt downward toward the west. The larger, easily visible clasts are nearly all fragments of felsic volcanic rock ranging in colour from dark maroon to brick red, orange, and pink. (Geologists have identified a likely source in similar volcanic rocks on the nearby mainland.)

Some clasts are crystal tuff speckled by white grains of feldspar and a dark grey variety of quartz. Others are banded rhyolite with well-preserved, millimetre-scale flow banding (see FYI). A small number of dark grey mafic pebbles are also part of the mixture.

Some groups of large clasts are strung along parallel to the rock layering (photo **b**), but they don't form continuous layers. Apart from these few groups, large and small pieces are jumbled together (photo **c**). Even the sand and gravel matrix itself is poorly sorted, with a range of grain sizes mixed in a featureless volume.

The range of colours and the detailed features of this attractive rock are best viewed in morning light or under a bright, overcast sky.

Rock Unit

New Mills Formation,
Dickie Cove Group

Bedrock Map

MP 2012-48

FYI

- The outcrop includes some freshly broken cobbles of banded rhyolite, making it easy to study this rock type closely. Look for large cobbles with millimetre-scale stripes in various shades of pink or orange.

  These colourful stripes are known as flow banding. After the volcanic eruption in which the rhyolite originally formed, the lava was a hot, syrupy mass like molten glass and could flow very slowly under its own weight. During that time the banding and its small-scale bends and buckles formed as the cooling lava oozed and sagged. A layer's colour depends partly on the extent to which the iron in it was oxidized during that time (the more oxidized layers are redder).

Banded rhyolite.

- The outcrop at West Point Island wharf is full of volcanic debris and probably formed near a volcanic vent, yet the rock is considered to be sedimentary, not igneous. Why?

  If volcanic debris is buried where it fell (for example, by more eruptions), the rock that forms from it is classified as igneous. But if volcanic debris is moved to other locations, whether by landslides or by melting snow- or rain-fed streams, then the rock that eventually forms—even though all its material is volcanic—is considered an epiclastic (sedimentary) rock, like this one. The key is whether or not it formed directly from volcanic processes.

Also Nearby

The rock at this site is part of a series of layers formed by an intense period of bimodal volcanic activity around 420 million years ago. Felsic and mafic volcanic rocks and related sedimentary rocks of this age extend inland from the coast between Charlo and the Jacquet River.

If you encounter rock outcrops on the east side of West Point Island or other sites nearby to the east, you may notice the rocks are dark shades of grey, green, or brown. They are mafic volcanic rocks—a fault cuts across West Point Island, and a different rock formation occupies the eastern half.

A former quarry, now filled by spring water, provides scenic and recreational enjoyment for visitors to Atlas Park in Pointe-Verte.

Old Ocean

A Slice of Gabbro from a Long-Lost Ocean Floor

The 1950s and 1960s brought a flurry of railway construction to northern New Brunswick, partly due to the discovery of valuable mineral deposits south of Bathurst (see site 38, Middle Landing). To lay new track, Canadian National Railway needed high-quality crushed stone to stabilize the railroad ties and support the weight of heavy, ore-laden trains.

It's no coincidence that Pointe-Verte's Atlas Park is beside the railway line. Its popular, scenic lake originated as a quarry established to supply crushed stone to the railway, making delivery easy. And the rock type here—gabbro (not limestone as reported on the park signs)—is prized in road and rail construction because in crushed form it is strong, packs stably, and drains well.

Thankfully, the quarry was converted to a park, so it's easy to view the gabbro, which has an exotic history. As part of an ophiolite complex, it preserves a slice of sea floor that formed during the Ordovician period. The sea is gone, most of its sea floor subducted deep into the Earth's mantle. Come see this rare remainder.

Getting There

Driving Directions

From Route 11 take Exit 333 and follow Rue de la Gare toward Route 134 and Pointe-Verte. About 4.5 kilometres from Route 11, just past the railroad track, watch for the park entrance (N47.84829 W65.76864), and turn right (south).

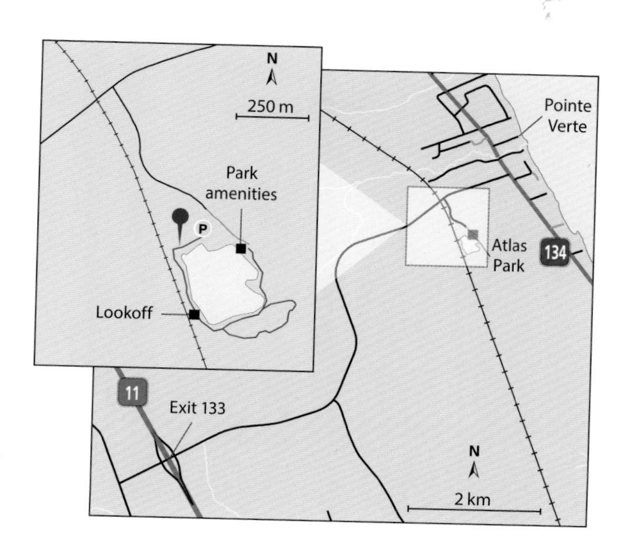

Or, from Route 134 in Pointe-Verte, find the intersection with Rue de la Gare (N47.85069 W65.75983) and follow it about 700 metres, then turn left into the park entrance.

Follow the park road about 450 metres to the main parking lot.

Where to Park

Parking Location: N47.84525 W65.76567

This large parking lot is located near the Visitor Information Centre and other park amenities.

Walking Directions

From the southwest corner of the parking lot, follow the gravel Hummingbird Trail for about 100 metres along the north side of the lake to the first outcrop location. Follow the trail around to the left (parallel to the railroad track) to see additional rock exposures.

Notes

Atlas Park has several trails, only two of which (Hummingbird and Owl Trails) are shown above, and other amenities. For more information about the park, visit http://parcatlaspark.ca.

1:50,000 Map

Pointe Verte 021P13

Provincial Scenic Route

Acadian Coastal Drive

On the Outcrop

The Hummingbird Trail in Atlas Park passes several large outcrops of gabbro like this one.

Outcrop Location: N47.84468 W65.76638

The rock outcrops along the Hummingbird Trail are mostly well weathered, obscuring the black and white mineral colours typical of gabbro. But several features are still visible. At the site pictured above, look for changes in the surface texture between coarse- and fine-grained areas of the outcrop. To the touch, these areas will feel like coarse and fine sandpaper.

In some outcrops, for example near the lookout by the railway line (N47.84334 W65.76584), you may notice well-defined dykes. Foot-worn outcrops may even allow you to see colour contrasts between dykes and their host rock.

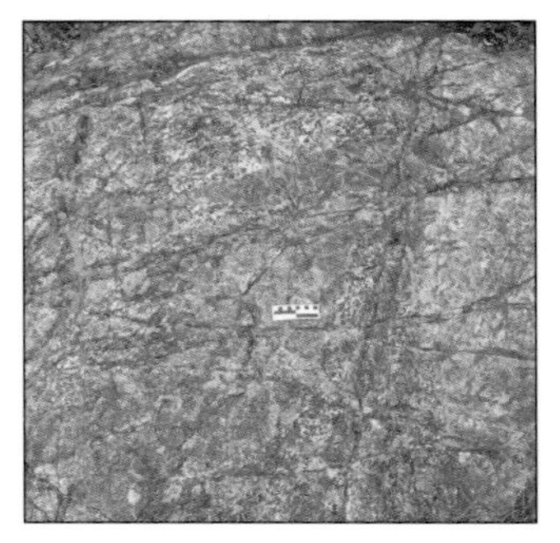

Gabbro.

One prominent feature typical of the outcrops has been accentuated by weathering. Their surface has a wrinkled "elephant skin" appearance, scored by a network of shallow grooves. These are narrow zones of intense deformation and mineral alteration. They formed as the ophiolite complex was being thrust onto adjacent continental rocks.

Rock Unit

Devereaux Complex

Bedrock Map

MP 2012-50

FYI

- The igneous rocks of an ophiolite complex (see diagram) form along a sea-floor ridge. There, slow but constant plate tectonic motion pulls the sea floor apart. Through the resulting fractures, molten rock flows from hot regions below and erupts onto the sea floor as pillow basalt. The basalt is gradually covered by fine mud and other deep-sea sediment.

  Below the basalt, magma left inside the fracture cools to become a solid dyke. Over time, a layer forms at that level with nothing but criss-crossing dykes (a sheeted dyke complex). The magma supply from below eventually cools to form the gabbro layer, as seen in Atlas Park. Beneath the gabbro are ultramafic rocks of the mantle.

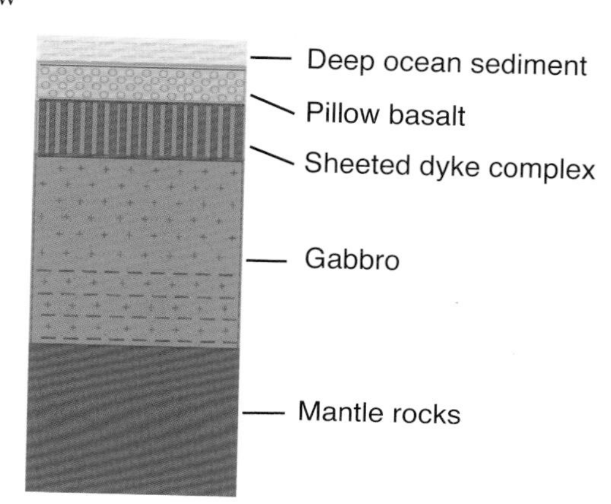

- The Ordovician sea floor where the Atlas Park gabbro formed is known by geologists as the Tetagouche back-arc basin. It was an ocean region situated between a volcanic arc and the continental margin of Ganderia—like today's Sea of Japan.
- The gabbro seen in the park is part of an area about 20 by 35 kilometres wide, known to geologists as the Elmtree inlier or Elmtree terrane. It includes other rocks typical of an ophiolite complex: pillow basalt, sheeted dykes, and deep-sea sediment, all formed during the Ordovician period.
- Later, during the Silurian period, as Ganderia collided with Laurentia, the layers of the Elmtree terrane's ophiolite complex were dissected by giant thrust faults and dealt out onto the adjacent continental crust like a deck of cards. The site of the thrusting is preserved by a highly deformed zone (mélange) containing fragments of altered mantle rock—a testament to the extreme forces involved.

Exploring Further

Oregon State University. "Ophiolites." *Volcano World.* http://volcano.oregonstate.edu/ophiolites.

The Nepisiguit River flows among thick slabs of granite at Pabineau Falls south of Bathurst.

Quarry Stone

Undeformed Devonian Granite

The Nepisiguit River is one of New Brunswick's most scenic, with numerous waterfalls, rapids, and rocky gorges. For more than 150 years it has prompted enraptured accounts by visiting naturalists and amateur fishers, including William Hickman,* who in 1860 described it as a place of "green trees, cool clear waters, salmon, rapids, canoes, and the other concomitants to this life of health and freedom."

For about 20 kilometres south of Bathurst, the Nepisiguit flows over a granite intrusion named for Pabineau Falls, where the rock is well exposed. At first glance the sheer walls of bedrock along the river may seem to be layered, like a sedimentary rock. But the Pabineau Falls granite is definitely igneous, part of a long episode of plutonic and volcanic activity that affected this region in Devonian times (see site 34, Inch Arran Point, FYI).

The appearance of layering is caused by nearly horizontal fractures known as exfoliation joints (see FYI). Thanks to such fractures, the rock proved convenient to quarry, providing an attractive, durable building stone. It has been sold under a variety of names, including Bathurst, Connolly, and Nepisiguit granite.

* No known relation to Martha.

Getting There

Driving Directions

From Route 11 in Bathurst take Exit 304 and follow Route 430 south for about 4.5 kilometres. Fork left (N47.55494 W65.67403) onto Pabineau Falls Road and follow it south for about 5.5 kilometres, passing through the Pabineau First Nation community along the way. Just before the paved portion of the road ends, watch for a pull-off on the right. The river is visible on the left.

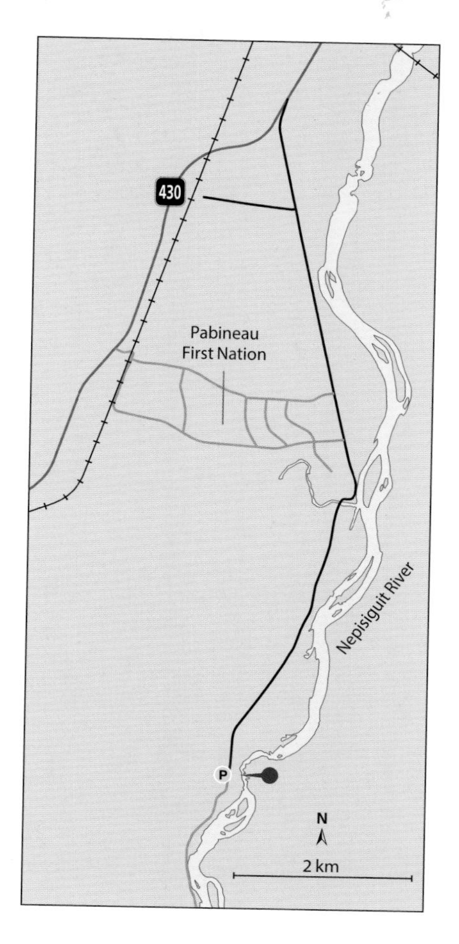

Where to Park

Parking Location: N47.50802 W65.67849

Park in the gravel pull-off on the right-hand side of the road.

Walking Directions

From the pull-off, walk across the road and follow a gravel lane toward the river. As conditions allow, approach the river and explore the rock pavement along the banks.

Notes

For more information about the Pabineau First Nation, visit https://www.pabineaufirstnation.ca.

This site is best visited in a dry season when river levels are low.

1:50,000 Map

Bathurst 021P12

Provincial Scenic Route

Acadian Coastal Drive

On the Outcrop

Near Pabineau Falls, extensive pavements of river-washed granite are exposed in the dry season.

Outcrop Location: N47.50790 W65.67741

During dry seasons when water levels are low, you can walk right onto a large area of granite at the outcrop location. Such a wide expanse allows you to appreciate the uniform, unblemished, even-grained quality of the stone. Some rock surfaces are especially pristine. They have been scrubbed by rushing, sediment-laden water in times of high flow.

This is a true granite, rich in pink potassium feldspar. On clean surfaces you will also see white plagioclase feldspar and light grey quartz. Less obvious are small grains of black biotite. The granite contains few variations of any kind. Small inclusions (pieces of contrasting rock) or veins of pegmatite (large crystals) or aplite (very small crystals, like sugar) are rare.

Pabineau Falls granite.

Together, these features imply that the rocks at this site formed near the centre of an intrusion that crystallized slowly and peacefully at depth after deformation in the region had ceased.

Rock Unit	**Bedrock Map**
Pabineau Falls Granite	MP 2013-5

FYI

- Granite commonly fractures parallel to its exposed surface, forming large, flat slabs like those seen along the Pabineau River. The slabs are called exfoliation sheets. The Pabineau granite was prized for quarrying partly because the exfoliation sheets were of useful thickness. They made it easy to remove large, workable blocks of stone with little drilling or waste.
- Exfoliation is a common process, but one that is difficult to account for. Stress caused by the intrusion process or later cooling, subtle tectonic forces, release of pressure as erosion brings rocks to the surface, and even heating and cooling by air and sun have all been considered as causes. It seems likely that several factors act in combination to cause exfoliation.

Also Nearby

For more than 120 years, until 1985, the Pabineau granite was quarried by the Connolly family of Bathurst. Initially used for bridge construction along the Intercolonial Railway, this granite also graces the city of Bathurst in the form of several prominent buildings.

A walking tour of about 1.4 kilometres (map at right) takes you past several fine examples. On St. Andrew Street are Sacred Heart Cathedral and related buildings (1, 2, and 3) as well as the fire station (4). Nearby on St. Patrick Street are the Gloucester County courthouse (5) and Canada Post office (6). A short digression down King Avenue to St. John Street brings you to Bathurst High School (7). From there, a walk along Coronation Park brings you back to the Cathedral.

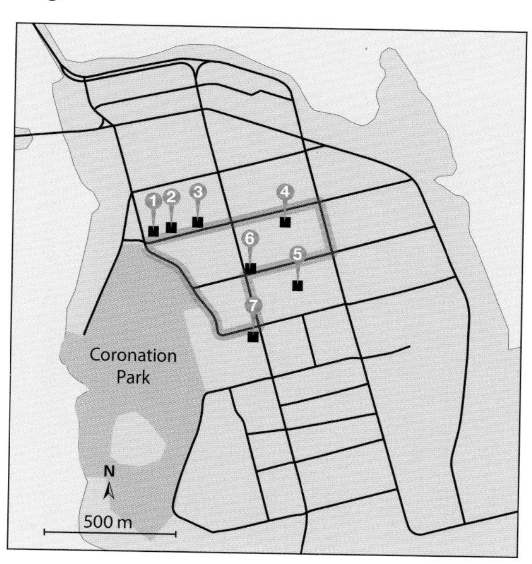

Exploring Further

Hickman, William. *Sketches on the Nipisaguit, a River of New Brunswick, B.N. America*. Halifax: John B. Strong, 1860. https://archive.org (search for "Sketches on the Nipisaguit").

Martin, Gwen. *For Love of Stone: The Story of New Brunswick's Building Stone Industry*, vol. 1, pp. 137–139. Fredericton: New Brunswick Department of Natural Resources and Energy, 1990. www1.gnb.ca/0078/geosciencedatabase (use PARIS search and enter title "for love of stone%").

At Middle Landing the Nepisiguit River runs through a narrow, rocky passage beside the Route 360 road bridge.

Below the Ore

Highly Deformed Ordovician Shale

Middle Landing has been popular with outdoor enthusiasts for more than 150 years. In 1860 a touring fly fisherman called it "a good and picturesque camping-ground." An 1876 report from Bathurst said that tourists were "pouring in from all parts of the country" to visit sites along the Nepisiguit River, including Pabineau Falls (site 37) and Middle Landing. Well-used trails and ATV tracks show it's still popular today.

Along with tourists came naturalists. After geologist Loring Bailey travelled down the river from Mount Carleton (see site 32, Williams Falls) in 1863, he predicted great mineral wealth would be found upstream from Middle Landing. In 1897, William Hussey of Bathurst found an iron deposit near Nepisiguit Falls. Tests near Hussey's find eventually led to the 1952 discovery of a valuable copper-lead-zinc deposit—and the prospecting rush that Bailey had foreseen was finally on.

The rocks at this location are on the doorstep, so to speak, of the mineral-rich Bathurst Mining Camp (see FYI). Here at Middle Landing are the rocks on which the ore-bearing layers were deposited—their foundation—formed early in the Ordovician period.

Getting There

Driving Directions

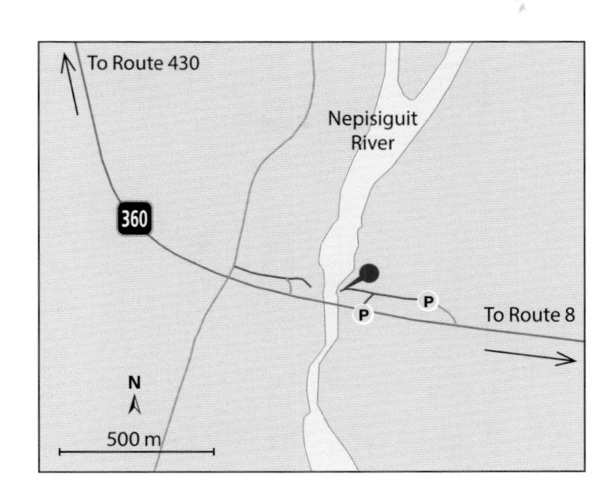

From the north, on Route 11 in Bathurst take Exit 304 and follow Route 430 south for about 16 kilometres to the intersection with Route 360 (N47.47317 W65.74906). Follow Route 360 about 5 kilometres to one of the parking locations by the Nepisiguit River.

From the east, on Route 8 take Exit 231 near Allardville and follow Route 360 west for about 14.5 kilometres to the site.

Where to Park

Parking Locations: (1) N47.44142 W65.70283
(2) N47.44120 W65.70462

This area is popular with outdoor enthusiasts and provides a number of possible parking areas, for example: (1) a clearing at the intersection of a gravel road and several trails or (2) on the shoulder just beyond the bridge guardrail on the south side of the road.

Walking Directions

A trail extends straight from parking area (**1**) toward the river, ending in a rock pavement beside the river. Follow the trail west about 200 metres to the river. (**2**) On the north side of the road, a trail emerges from the woods near the bridge. Follow this trail about 35 metres to an intersecting trail and turn left. Walk a further 70 metres to the river.

As conditions allow, explore the rock pavement beside the river.

Notes

This site is best visited in a dry season when river levels are low.

1:50,000 Map

Nepisiguit Falls 021P05

Provincial Scenic Route

Acadian Coastal Drive

On the Outcrop

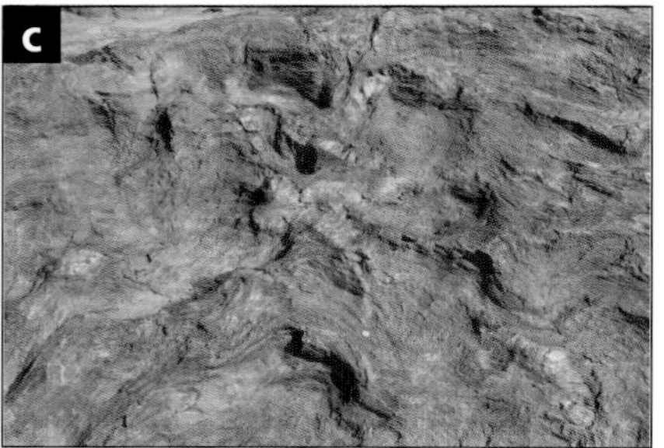

At Middle Landing, intensely deformed and metamorphosed black shale encloses a variety of less-deformed rock types: (**a**) siltstone, and (**b** and **c**) deformed and fragmented quartz veins.

Outcrop Location: N47.44161 W65.70526

Geologist R.W. Ells passed along this stretch of the river in 1881 and, with a gift for understatement, declared the rocks "a good deal disturbed." Certainly, the main impression they give is one of severe deformation. But these rocks originated as flat-lying layers of black shale and a type of "dirty" siltstone known as quartz wacke (silt containing bits of feldspar and clay as well as quartz). Shades of orange in the rock are due to the presence of iron sulphide, which rusts.

Metamorphism and folding converted the original shale to slate, with millimetre-scale cleavage. When initially formed, the cleavage must have been aligned along even, parallel planes (for example, see site 31, Edmundston). But later episodes of deformation crumpled and folded it, as you can see on some of the rock surfaces. Because metamorphosed siltstone is relatively rigid and resists deformation, it survives as large, broken fragments surrounded by the contorted slate (photo **a**).

In the outcrop you may also see folded, broken veins of quartz (photos **b** and **c**). These likely formed during metamorphism, then were affected by later episodes of deformation.

Rock Unit

Patrick Brook Formation,
Miramichi Group

Bedrock Map

MP 2014-3

FYI

- Geologists see evidence of four or five separate deformation events in the rocks along this part of the Nepisiguit River. Deformation resulted from a series of terrane collisions in the Ordovician, Silurian, and Devonian periods.
- Although deformation is the most obvious feature here, the early history of these rocks is also interesting. Just 5 kilometres upstream from Middle Landing, rocks at the Chain of Rocks rapids originated as Cambrian quartz sandstone and greywacke layers on the continental margin of Ganderia. The Ordovician black shale at Middle Landing was deposited on them. It is a well-known sequence throughout the terranes of Ganderia (see site 29, Hays Falls, FYI).

Also Nearby

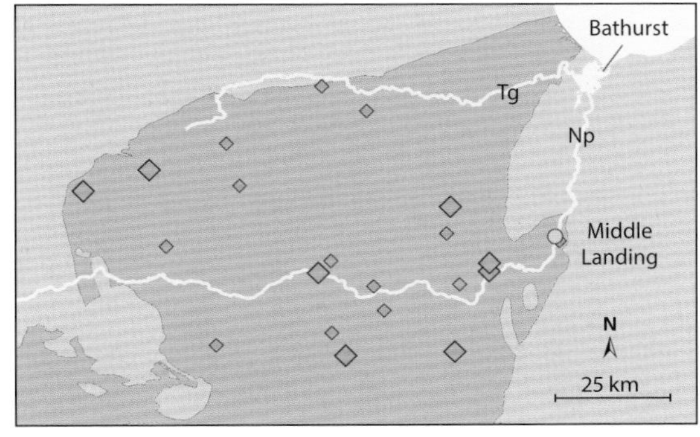

The Bathurst Mining Camp is world-famous among geologists for its numerous ore deposits. It encompasses an area about 70 kilometres wide, much of it between the Tetagouche (Tg) and Nepisiguit (Np) Rivers (see map at right). The map shows former mine sites (large symbols) and a sampling of other significant deposits (small symbols), all within an area of Cambrian and Ordovician rocks (shown in green) southwest of Bathurst (younger rocks, grey; Middle Landing, yellow dot).

The mineral deposits formed as volcanogenic massive sulphide deposits (known as VMS deposits) in the Tetagouche back-arc basin (see site 36, Atlas Park, FYI). Underwater volcanic activity caused very hot, mineral-rich fluids to circulate for millions of years among the sedimentary and volcanic rocks on the basin floor. Stable, oxygen-poor conditions in the sea water allowed the sulphur-based minerals to accumulate rather than dissolve.

To date nearly 200 million tonnes of mineral ore containing copper, lead, zinc, precious metals, and other valuable materials have been extracted by operations in the Bathurst Mining Camp.

Exploring Further

Woods Hole Oceanographic Institute. "Up in 'Smoke.'" *WHOI in Motion*. www.whoi.edu/VideoGallery (click on the video's icon to see an underwater, mineral-rich fluid in action).

Sandstone at Cap-Bateau is dark reddish brown, more like the rocks of Prince Edward Island than those farther west in New Brunswick.

Go East

A Colour Change Late in the Carboniferous Period

The islands of Lamèque and Miscou reach out from Shippigan on the Acadian Peninsula into the Gulf of St. Lawrence. You can drive there, thanks to two long causeways on Route 113. But you'll know you've entered an island world—big water, big sky. A change in the air alerts you that you've left the mainland.

The rock outcrops along New Brunswick's easternmost shores are among the youngest in the Maritimes basin of the province, deposited near the end of the Carboniferous period. Along the scenic journey to this site, you can explore the Acadian Peninsula's heritage of sandstone, used both in industry (see Also Nearby) and architecture (site 40, Grande-Anse to Neguac).

At the time the rocks at Cap-Bateau formed, fault movements had caused uplift in what is now the Gaspé Peninsula to the north. Sandstone deposited there earlier in the Carboniferous period was being eroded—basically, recycled—as rivers carried the sand to this area. The previously humid tropical climate of the Coal Age started to dry out, bringing big changes to the region as Pangaea's assembly progressed.

Getting There

Driving Directions

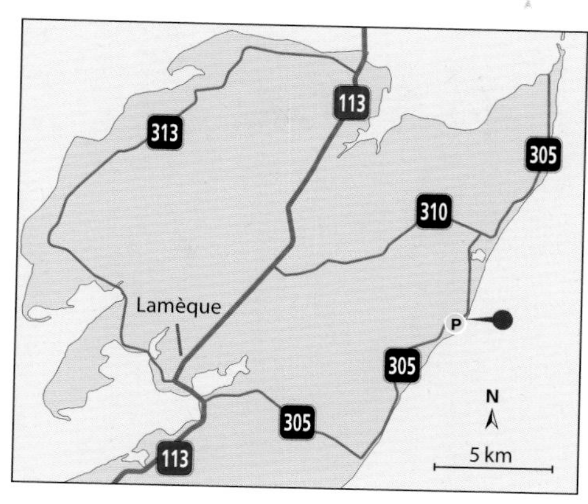

From Route 11 about 14 kilometres south of Caraquet, turn onto Route 113 (N47.67594 W64.87905). Follow it east then northeast for about 25 kilometres, passing through Shippigan and onto Lamèque Island. In the community of Lamèque, just before you reach the causeway turn right (east) onto Route 305 (N47.78806 W64.63628). Follow Route 305 east to the coast, then northeast along the coast to Cap-Bateau. The trip along Route 305 is about 11 kilometres in all from Lamèque. As the road approaches the open shore on a curve in Cap-Bateau, watch for a sandy lane on the right, which leads to the parking location.

Where to Park

Parking Location: N47.81192 W64.53751

This open sandy area about 50 metres from the road lies at the juncture of several trails.

Walking Directions

From the parking area, walk onto the sand and gravel beach and as conditions allow, approach the rock outcrops on your left (north). If the tide is low, you may be able to explore outcrops on the south end of the beach.

Notes

This site is best visited in quiet, low-tide conditions. About 4.5 kilometres east of Route 11, Route 113 also passes through Inkerman, the site of one of the Acadian Peninsula's fine stone churches (see site 40, Grande-Anse to Neguac).

1:50,000 Map

Caraquet 021P15

Provincial Scenic Route

Acadian Coastal Drive

On the Outcrop

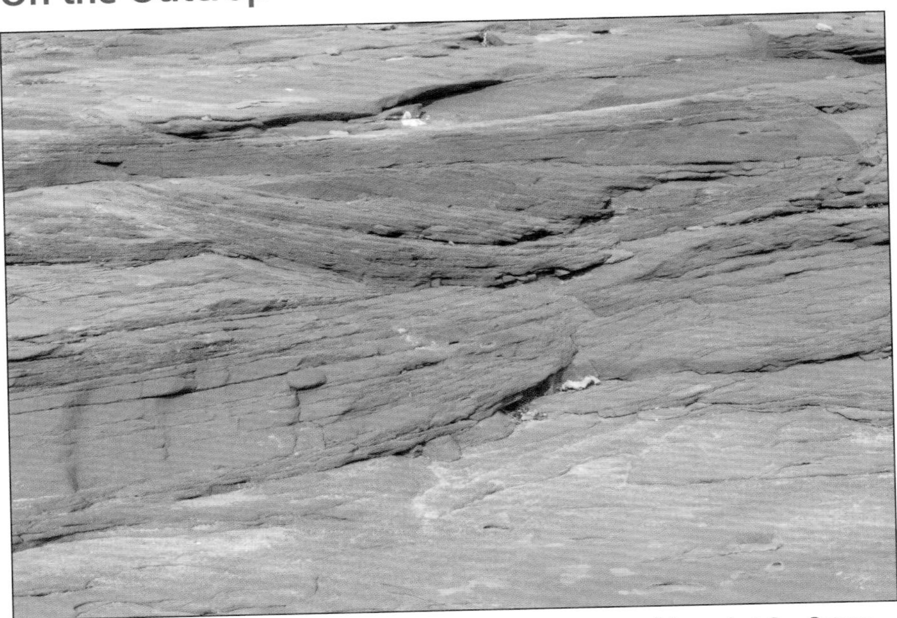

Tilted laminations known as cross-bedding attest to the river origins of the rock at Cap-Bateau.

Outcrop Location: N47.81165 W64.53672

The rock at this site is reddish brown, river-deposited sandstone. It is not strongly cemented and weathers easily into softly sculpted shapes. If you brush the surface with your hand, you may even feel loose grains on the surface. This property of the rock has contributed to the long sandy beaches typical of New Brunswick's eastern shores.

Weathering has also made it easy to see interesting features on the rock surface. Notice that the layering does not follow a regular, horizontal pattern. Some groups of thin layers form lentil- or lens-shaped areas of slightly curved, tilted laminations. These formed along the sides of water channels, sandbars, and other riverbed features as sediment-laden water flowed past (see site 2, St. Andrews, FYI).

Former sea arch.

The physical properties of the rock at Cap-Bateau mean that the coastline is subject to rapid change. Until a few years ago along this stretch of shoreline was a sea arch (photo at right). Coastal erosion has since demolished it.

Rock Unit

Clifton Formation, Pictou Group

Bedrock Map

NR 1 (2008)

FYI

- Throughout the Carboniferous period, the climate of this region changed from arid to humid (during the Coal Age) and back to arid again. This trend is recorded by numerous changes in sedimentary rocks, including their colour, in various shades of red, grey, or brown. Red varieties of sandstone typically form in an arid climate, as was the case near the end of the Carboniferous period when the rocks at Cap-Bateau formed.

Also Nearby

If you are travelling between Bathurst and Cap-Bateau, consider visiting additional sandstone sites along Chaleur Bay (map below) or some of the sandstone churches in the area (site 40, Grande-Anse to Neguac).

In the past, sandstone was prized for its usefulness. A variety from Stonehaven was fashioned into grindstones. Near the public wharf in Stonehaven (N47.75399 W65.36190), you can see the former quarry location.

Public beaches at Grande-Anse (N47.82275 W65.14362), Pokeshaw (N47.78930 W65.25112), and Caraquet (N47.79448 W64.94120) all offer views of adjacent sandstone cliffs. All three areas were sources of building stone.

With the advent of carborundum for grinding and of structural steel and concrete for building construction, by 1930 the need for natural sandstone was virtually gone, and local knowledge of stonecraft gradually faded.

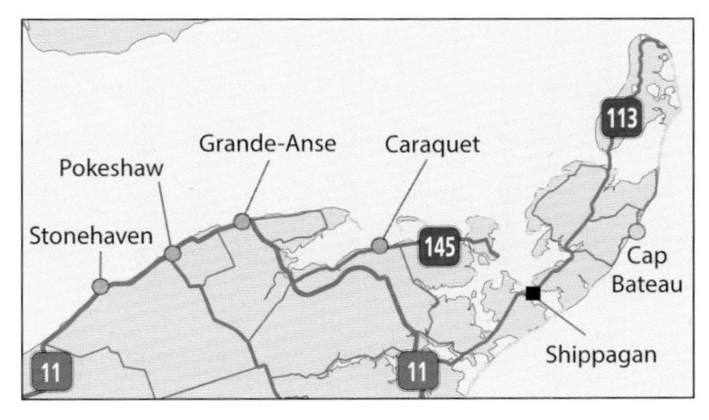

Sandstone cliffs, Stonehaven.

Exploring Further

New Brunswick Museum. *Magnificent Rocks.* "Gallery Search." http://www.magnificentrocks-rochesmagnifique.ca (under Search for Images, enter "Stonehaven" for photos of Stonehaven quarry and quarry products).

800 700 600 500 400 300 200 100 0
Z1 Z2 Z3 Є O S D C P Ꞓ J K Cz

The sandstone church at Grande-Anse stands near the cliffs from which the rock was quarried.

Stone Tour

The Acadian Peninsula's Sandstone Heritage

As you travel the gently rolling landscape of eastern New Brunswick, its Carboniferous foundation of sandstone is mostly hidden. With outcrops limited to river valleys and stretches of shoreline, sandstone is everywhere, yet nowhere. On the Acadian Peninsula, though, geological and human history have combined to reveal this stone abundantly—in the form of churches.

Stonecraft is deeply woven into Acadian history. Acadian settlers quarried stone for tool sharpening, grain milling, and building construction. From about 1800 to 1930, operations at nearby commercial quarries (see site 39, Cap-Bateau, FYI) also helped hone and perpetuate local skills. Because of this widespread knowledge, local labour and expertise for the construction of stone churches were available even in small communities.

In consequence the Acadian Peninsula is home to the largest concentration of sandstone churches anywhere in New Brunswick. The area between Grande-Anse and Neguac includes eight impressive examples. Most of the building stones in the churches' exterior walls were quarried from thick, homogeneous layers of sandstone. The well-sorted sand in these rocks was derived from the erosion of older sandstone (see site 41, French Fort Cove).

Getting There

Driving Directions

Between Bathurst and Miramichi, Route 11 makes a big loop around the Acadian Peninsula, converging with Route 8 at each end. Some of the stone churches are on Route 11 itself; the rest are within about 10 kilometres of Route 11 along other numbered routes. All have plentiful parking on site.

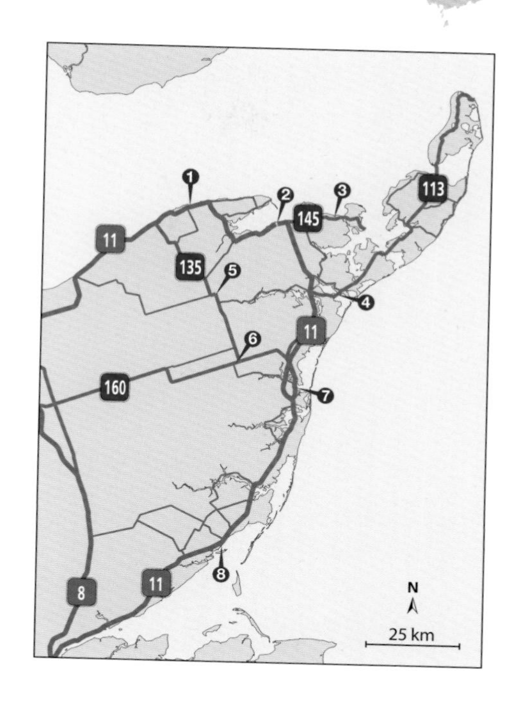

As an example, the churches could be visited in the order shown on the map with a total driving distance of about 150 kilometres and a driving time of about two hours in ideal conditions.

1. St-Simon et St-Jude, Route 11, Grande-Anse
2. Saint-Pierre-aux-Liens, Route 11, Caraquet
3. Saint-Paul, Route 145, Bas-Caraquet
4. Saint-Michel, Rue de l'Église (intersects Route 113), Inkerman
5. Saint-Augustin, Route 135, Paquetville
6. Saint-Isidore, Route 160, Saint-Isidore
7. St-Jean-Baptiste et St-Joseph, Rue Principal (intersects Route 11; take Exit 194), Tracadie-Sheila
8. Saint-Bernard, Route 11, Neguac

1:50,000 Map

021P06, -07, -10, -11, -14, -15

Provincial Scenic Route

Acadian Coastal Drive

1. Grande-Anse

This church was constructed in 1947–1949 of stone salvaged from an earlier church on this site, dating from 1902 but destroyed by fire in 1946.

The Grande-Anse quarry was located on the shore about 500 metres west of the church. Local residents transported stone to the site by horse-drawn sled over the frozen waters of Chaleur Bay in the winter of 1902.

The head stonemason for the original 1902 church had learned his trade in the nearby Stonehaven grindstone quarry.

St-Simon et St-Jude.

Location: N47.81544 W65.17546

2. Caraquet

This is the oldest surviving stone church on the Acadian Peninsula, and one of the oldest in New Brunswick. It was built in 1857–1864.

Most of the stone was quarried in Caraquet and Grande-Anse. A few of the more weathered blocks are thought to have been recycled from an earlier church on this site, dating from 1817.

Beside the church are the remains of a convent also constructed of stone recycled from the previous church. Stone for a 1905 addition to the convent was likely quarried locally.

Saint-Pierre-aux-Liens.

Location: N47.78984 W64.96475

Rock Unit

Clifton Formation, Pictou Group

Bedrock Map

NR 1 (2008)

3. Bas-Caraquet

Stone for this church was brought in via Chaleur Bay from quarries in Caraquet and Grande-Anse. Horse-drawn sleds hauled loads of stone over the ice-covered bay in winter; schooners made the trip in summer. Over a period of two years, local parishioners, joined by the parish priest, extracted and moved the stone.

The church was built in 1902–1905 under the direction of a head mason from Saint-Isidore (stop 6). The buttresses were refurbished in 1988, so parts of them appear less weathered.

Saint-Paul.

Location: N47.80094 W64.82977

4. Inkerman

Parishioners extracted stone for this church from two quarries along the Pokemouche River west of Inkerman, carting a few blocks per day to the site.

The church was completed in two years, 1916–1918, under the direction of the master stonemason who had overseen construction of the church in Grande-Anse (stop 1) in 1902.

This structure is unique in the region for its two-tone construction in red and golden brown sandstone. The small size of the red stone blocks may have been necessary due to the scale of the layering in the original sandstone.

Saint-Michel.

Location: N47.67145 W64.81951

Rock Unit

Clifton Formation, Pictou Group

Bedrock Map

NR 1 (2008)

5. Paquetville

Parishioners quarried stone for this church from two sites, one about 1 kilometre north of town and the other in St. Amateur, about 5 kilometres to the west. Construction was completed in 1920–1928 by stonemasons who had learned their trade in the Stonehaven grindstone quarry, including the same master mason who had overseen construction of churches at Grande-Anse and Inkerman (stops 1 and 4).

The sandstone is noticeably coarser here than at other stops. Few sedimentary features are visible in the homogeneous, weather-resistant stone.

Saint-Augustin.

Location: N47.66845 W65.10844

6. Saint-Isidore

Horse-drawn sleds hauled stone to this site from a quarry along the Big Tracadie River. Construction was completed by local masons in 1904–1908, under the direction of a master mason from Quebec.

The sandstone here is finer grained than the rock used in Paquetville but is similarly massive and homogeneous. The stone has resisted weathering, with sharp, clean lines surviving in exterior carved features.

Designated a provincial historic site, the church has an ornately finished, white-and-gold interior.

Saint-Isidore.

Location: N47.55298 W65.05647

Rock Unit

Clifton Formation, Pictou Group

Bedrock Map

NR 1 (2008)

7. Tracadie-Sheila

Tracadie, like Grande-Anse and Caraquet, had a history of commercial stone production. A stone church dating from 1873 at this site was destroyed by fire in 1925. Its replacement was built on the surviving foundation beginning the following year.

Some of the stone here is rather coarse grained, with visible pebbles up to about 3 or 4 millimetres across. However, overall it is homogeneous.

Notice the flat, pecked finish of the exterior walls, in contrast to the hewn finish at other stops.

St-Jean-Baptiste et St-Joseph.

Location: N47.51360 W64.91178

8. Neguac

Neguac, settled by displaced Acadians during the Great Deportations, is steeped in Acadian history. Nearby is a designated historic place, the homestead of early resident and Saint-Bernard parish warden Otho Robichard.

Construction of this church extended from 1910 to 1914. It was later restored in 1946 following a fire.

Although the specific origin of the building stone is unclear, this golden brown, homogeneous sandstone is typical of the region (see site 41, French Fort Cove) and was almost certainly derived locally.

Saint-Bernard.

Location: N47.24495 W65.08076

800 700 600 500 400 300 200 100 0

Z1 Z2 Z3 Є O S D C P Ŧ J K Cz

At French Fort Cove in Miramichi, a day park with amenities provides access to several wooded trails and the remains of a historic quarry.

Quarry Tales

Carboniferous Recycling of Sand

"Charles E. Fish—the proprietor of French Fort Cove quarry in [Miramichi]—was beyond doubt or exaggeration the Don Quixote of the New Brunswick stone industry." So begins geoheritage author Gwen Martin's account of the history of quarry operations at French Fort Cove. Fish operated the quarry here from 1884 to 1904. Disorganized but ambitious, he supplied stone for the Langevin Block in Ottawa and St. Dunstan's Basilica in Charlottetown (see site 44), among others.

Fish succeeded despite his personal shortcomings in part because the rock itself has such desirable qualities. The sandstone's thick and extensive, homogeneous light brown layers owe their virtues to conditions during the Carboniferous period. Uplift along a large-scale fault within the Maritimes basin had exposed an adjacent area of older sandstone to erosion.

Rivers fed by abundant seasonal rainfall washed volumes of well-sorted, recycled sand down into the lowlands. It was deposited on flood plains and in large delta channels, forming the layers seen here today. As you walk the park's trails, enjoy the intermingled histories of the rocks and the quarry.

Getting There

Driving Directions

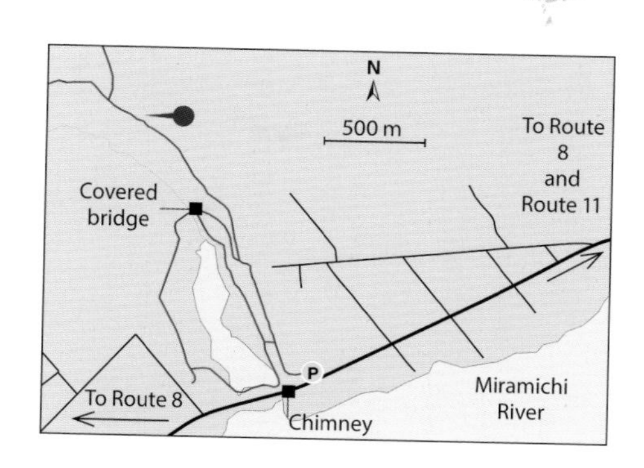

The King George Highway in Miramichi runs along the north side of the river. East of the Centennial Bridge it is part of Route 11. At the Centennial Bridge and the Miramichi Bridge it intersects Route 8.

Along the King George Highway from either direction, watch for the entrance to French Fort Cove (N47.01730 W65.54483) on the north side of the road. The entrance is located beside a large car dealership, about 5 kilometres west of the Centennial Bridge and about 4 kilometres east of the Miramichi Bridge.

Where to Park

Parking Location: N47.01729 W65.54525

This location includes a playground, picnic facilities, and other amenities. Alternatively, for closer access to the old quarry, limited parking is available at the end of French Fort Road (N47.02069 W65.54778).

Walking Directions

Maps of the trail system are posted at the trail heads, and key trail intersections are marked. Follow the signs for the Fish Quarry Loop. From the parking location it is about 400 metres to French Fort Road. Continue a further 850 metres (to the fork at Creaghan Gulch Loop, N47.02568 W65.55444) for abundant views of the sandstone.

Note: The map above does not show the complete Fish Quarry Loop, just that portion related to the text.

Notes

Along King George Highway near the park you may see a tall, round chimney near the mouth of the cove. It marks the site of Buckley's Mill, a sawmill that burned in 1922.

1:50,000 Map

Sevogle 021P04

Provincial Scenic Route

Acadian Coastal Drive

On the Outcrop

A former quarry face can be seen in places along the uphill side of the Fish Quarry Loop in French Fort Cove park.

Outcrop Location: N47.02541 W65.55318

The waypoint above marks a particularly large expanse of sandstone along the Fish Quarry Loop. There are numerous other exposures for 700 to 800 metres along the trail between French Fort Road and Creaghan's Loop.

Some of the sandstone layers are more than 1 metre thick (see photo at right; walking stick for scale) and homogenous in texture and grain size across the entire outcrop. These were best suited for building stone. Some of the thinner layers could have been used for making abrasive stones for sharpening and grinding. Even rubble had its uses in railway construction. The quarry delivered a diverse line of products.

Thick sandstone bed.

As seen in the photo above, the quarry site is becoming overgrown with trees. The signs of its industrial past are gradually fading, but with sharp eyes you can spot tool marks in some rocks along the trail (see FYI).

Rock Unit

Clifton Formation, Pictou Group

Bedrock Map

NR 1 (2008)

FYI

- At the intersection with Creaghan Gulch Loop, Fish Quarry Loop forks left. In the first 350 metres or so past the fork, blocks of leftover quarried material contain signs of the work methods that were used there.

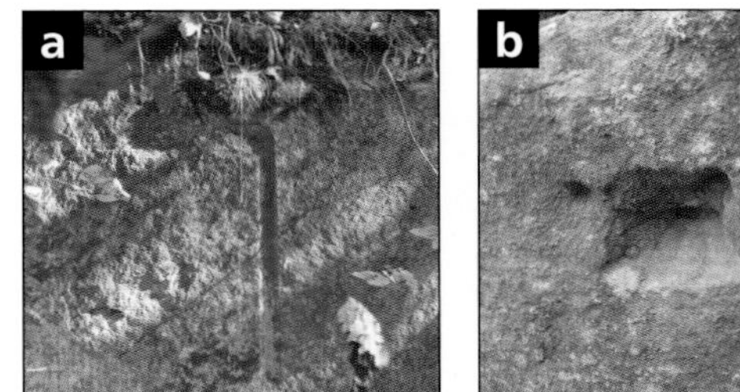

Signs of traditional quarrying methods along the trail include (**a**) a drill mark, along which stone was split for removal; (**b**) a chiselled mark, perhaps intended for a splitting wedge; and (**c**) a rough, chisel-dressed finish.

- Sandstone is porous and holds water while underground. When first quarried, the stone is soft and easily worked. Over time, moisture escapes through the exposed surfaces, leaving minerals behind that help harden the surface. This process, known as case hardening, is what makes sandstone durable as a building stone.

Also Nearby

If you are travelling Route 11 between Miramichi and Shediac, consider a detour to Cap Lumière. Where Route 505 makes a sharp turn, look for a little picnic park between the road and the shore (N46.65937 W64.70873). From there, as conditions allow, you can access the rocky shoreline.

Cap Lumière.

Here, the brownish grey sandstone of the Richibucto Formation is strongly cross-bedded, with one set of diagonal laminations stacked above another. This is best seen looking back toward the picnic park. In addition to quartz sand, this rock contains a lot of muscovite, a type of mica. Look for sparkling flakes of it in the rock surface.

Exploring Further

Martin, Gwen. *For Love of Stone: The Story of New Brunswick's Building Stone Industry*, vol. 1, pp. 90–96. Fredericton: New Brunswick Department of Natural Resources and Energy, 1990. www1.gnb.ca/0078/geosciencedatabase (use PARIS search and enter title "for love of stone%").

Colourful rock outcrops, a historic lighthouse, a wind farm, trails, and an interpretive centre are all attractions at North Cape, Prince Edward Island.

Drying Out

Sandstone from a Time of Climate Change

The North Cape lighthouse is one of the oldest in the province and a designated national historic place. Its light warns of a long rocky reef—visible only at low tide—that extends from the tip of the cape. The reef forms a boundary between the Northumberland Strait on the cape's western shore and the Gulf of St. Lawrence to the east.

The rock layers at North Cape mark a different sort of boundary—visible only in the mind's eye—between the Carboniferous and Permian periods of geologic time. Back then, the ocean breezes that turn the windmills of North Cape these days were nowhere to be found. The region sat near the equator, surrounded by land in all directions.

The rocks at North Cape formed at the beginning of the Permian period. The climate was becoming hotter, with long dry spells. During seasonal rains, runoff flowed quickly at times, allowing streams to move pebbles as well as sand and silt. One rainy season at a time, such streams built up the layers of conglomerate, sandstone, and siltstone preserved in the rocks of the cape.

Getting There

Driving Directions

From the intersection of Route 2 and Route 14 in Tignish (Phillip and Church Streets; N46.94999 W64.03298), travel north about 50 metres on Route 14, then take the first right at Dalton Avenue. Drive east on Dalton Avenue about 1.4 kilometres; the road ends at Route 12 (N46.95372 W64.01519). Turn left (north) and follow Route 12 about 12.5 kilometres to its end point at the North Cape wind farm.

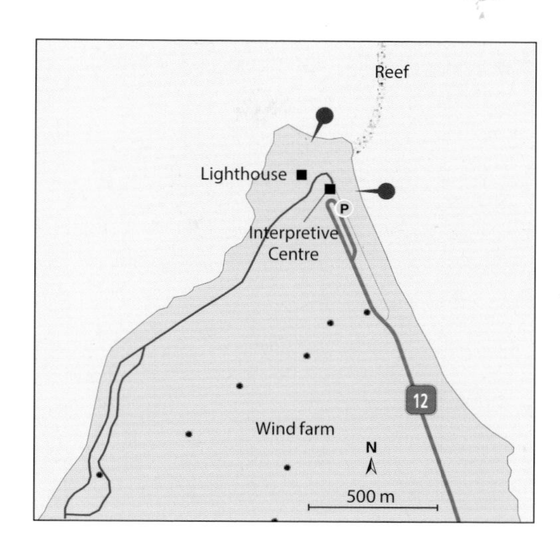

Where to Park

Parking Location: N47.05688 W63.99545

Plenty of parking is available in front of the Wind Energy Interpretive Centre.

Walking Directions

From the north end of the parking lot, facing the Interpretive Centre, turn right and walk toward the shore onto an unpaved lane beside the parking lot. Follow the lane north past the Interpretive Centre. Near the tip of the cape, numerous footpaths lead down the bluff onto the shore. Make your way to the shore at the tip of the cape (just west of the rock reef) and look back toward the lighthouse to view the low rock cliffs. Or, along the east side of the cape, double back toward the Interpretive Centre to explore rock pavements on the shore.

Notes

This site is best viewed in quiet, low-tide conditions. The park includes an extensive trail system, and several amenities are housed in the Interpretive Centre. Please stay on established footpaths to access the shore.

1:50,000 Map

North Cape 021P01/011M04

Provincial Scenic Route

North Cape Coastal Drive

On the Outcrop

This section (about 1 metre high) in the rock cliff at the tip of North Cape illustrates some of the colourful features typical of river sediments.

Outcrop Location: N47.05851 W63.99661

Prominent features in the cliff faces* include pods or lenses of conglomerate, which were deposited by quickly flowing water in a river channel. Areas of cross-bedded sandstone were deposited along the edges of a channel or sandbar. Sections of flat-lying, parallel layers were deposited in standing pools, perhaps after a flood. The red colour, small scale, and variability of the features suggest a generally dry landscape with occasional rainfall.

In some exposures you may see thin layers or wisps of light greenish grey sandstone. These sediments were starved of oxygen after burial, perhaps due to decaying organic material. This caused grey ferrous oxide to form instead of the red ferric oxide that colours the surrounding rock.

Along the shore on the east side of the cape (N47.05728 W63.99508) are rock pavements of conglomerate and coarse sandstone. Many of the conglomerate's pebbles are white quartz, but some are fine-grained volcanic rock in various colours.

* As storms erode the shore, the details of what you see may vary from those in the photo.

Rock Unit	**Bedrock Map**
Kildare Capes Formation, Pictou Group	Not available

FYI

- Geologists can use cross-bedding and other rock features to determine the direction of water flow in ancient river systems. Based on such evidence, the Permian rivers of this region were flowing from what is now central New Brunswick. That means the conglomerate pebbles at North Cape may have come, for example, from an area of Devonian igneous activity farther west (see site 32, Williams Falls, or site 34, Inch Arran Point, FYI).
- The sedimentary rocks of Prince Edward Island were deposited over a period of about 20 million years. The oldest rock formations, exposed in the western part of the island (1 and 2 in the map below), formed just before the end of the Carboniferous period. The rocks at North Cape (3) began to form when the Permian period began. Younger sedimentary layers (4 and 5) accumulated during the first part of the Permian period.

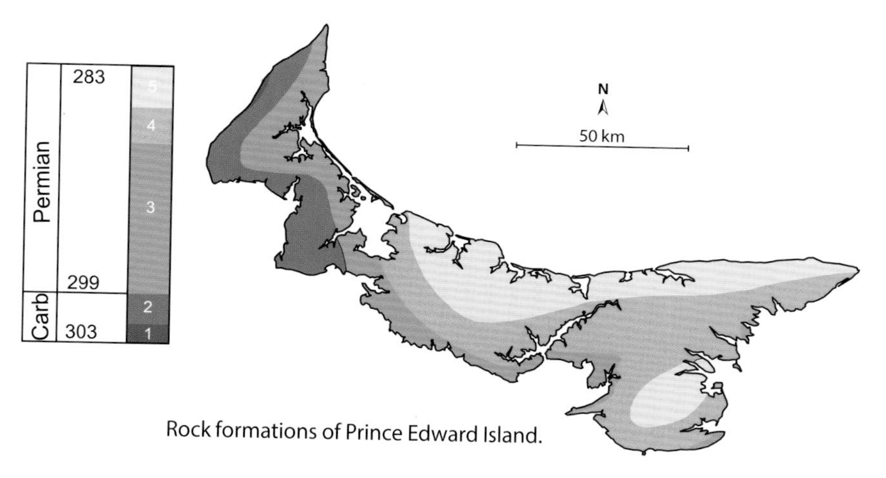

Rock formations of Prince Edward Island.

- Permian climate change was triggered by various factors, such as the final collapse and retreat of south polar ice sheets, which had moderated the Earth's tropical regions for millions of years. Plate tectonics also played a part. As the assembly of Pangaea progressed, more and more of the Earth's land became isolated from the open ocean, giving way to drier, harsher, continental conditions.

Exploring Further

Calder, John H. *Island at the Centre of the World: The Geological Heritage of Prince Edward Island.* Charlottetown: Acorn Press, 2018. An exploration of Prince Edward Island's geoheritage.

Red sandstone bedrock of Permian age is exposed along the shore beside Cavendish Beach in Prince Edward Island National Park.

C

Sparkling Sands

Prince Edward Island's Youngest Sandstone

Prince Edward Island National Park extends eastward from Cavendish Beach for more than 40 kilometres along the island's north shore. Over much of its length the park area consists of dunes, beaches, and wetlands, but at this site and others nearby sandstone bedrock is exposed along the shore.

Mention Cavendish and many will think of the long-popular novel *Anne of Green Gables* and its fictional community, Avonlea, both of which are also celebrated in the park. Partly based on the novel's legacy and related tourist initiatives, the island's red rocks seem synonymous with its fertility and rich agricultural heritage. But the world in which they formed could hardly have been more different from the present.

During much of the previous Carboniferous period, the landscape had been dominated by mighty rivers, swamps, and deltas. But the rocks here formed partway through the Permian period, and by then the region had become quite dry, like the arid tropical savannah of East Africa today. Rivers flowed only during seasonal rains. Plant diversity was in decline, resulting in an increasingly barren landscape as Earth headed toward climate disaster (see FYI).

Getting There

Driving Directions

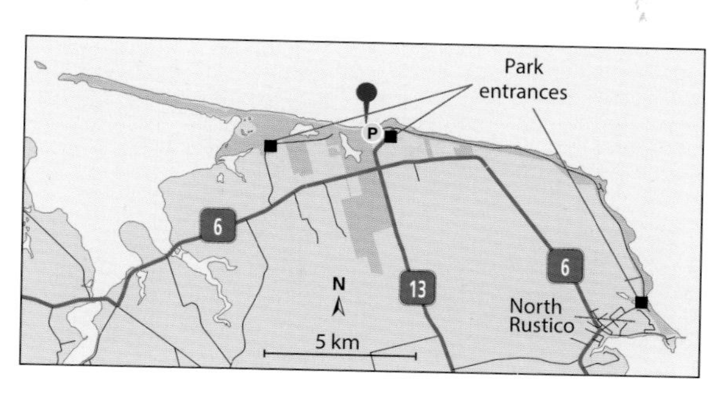

From the intersection of Route 13 and Route 6 (N46.49131 W63.37862) in Cavendish, follow Route 13 north to an entrance kiosk for Prince Edward Island National Park. Within the park, continue along Route 13 for about 200 metres then, following the signs for Oceanview, turn left onto Terre Rouge Lane (N46.49957 W63.37442). Follow the lane for about 500 metres. It ends in a large paved parking lot near the site.

Where to Park

Parking Location: N46.49940 W63.38114

Park in the paved lot at the west end of Terre Rouge Lane.

Walking Directions

From the northwest corner of the parking lot (on the right as you face the display board), follow a gravel trail west about 120 metres to a lookout. On the way back from the lookout, as conditions allow, follow footpaths across the grass onto adjacent rock pavements for a closer look.

Notes

This site lies within the boundaries of Prince Edward Island National Park and requires a valid park pass. Some outcrops here can be viewed and explored above the beach, so low tide is not required. Exploration at beach level, however, is best undertaken in quiet, low-tide conditions.

In addition to the parking location, there are several other pull-offs along the park's Gulfshore Parkway offering cliff views or beach access.

1:50,000 Map

North Rustico 011L06/011L11

Provincial Scenic Route

Green Gables Shore

On the Outcrop

The sandstone pictured here is characterized by thin, flat, parallel layers just a few millimetres thick. It weathers along these layers because the rock contains so much mica.

Outcrop Location: N46.49958 W63.38335

Much of the sandstone along this part of the coast contains visible flakes of shiny but colourless mica known as muscovite. This suggests that the sediment in the rocks was derived from a region rich in metamorphic or perhaps granitic rocks. The muscovite is easy to identify as brightly sparkling mineral flakes on the rock surface.

Mica forms flat flakes because its crystals form as stacks of weakly bonded sheets. In sedimentary rocks, mica flakes tend to be aligned parallel to the layering. At this site, concentrations of mica have emphasized fine layering in the sandstone, as seen in the photo above.

Cross-bedding visible in some vertical rock faces seen from the trails is evidence for the river origins of the sediments that formed these rocks. Weathering has accentuated the cross-bedding, which likely formed as part of a sandbar or channel deposit.

Cross-bedding.

Rock Unit

Orby Head Formation, Pictou Group

Bedrock Map

Not available

FYI

- These brightly coloured sedimentary rocks are known as red beds. In Prince Edward Island they formed due to the hot, dry conditions of Pangaea's interior. The sediment grains are not red themselves but rather are coated in a thin layer of iron oxide. See Geology Basics (Rock Types) for details.

  The sand of Cavendish Beach was derived from erosion of the sandstone, but is not very red. As wind and waves of the Gulf of St. Lawrence tumble the loose sand grains against one another year after year, the red coating is rubbed off.

- The rocks exposed in Prince Edward Island National Park between Cavendish Beach and North Rustico are among the youngest rocks of the Maritimes basin exposed on land. Only those on the Magdalen Islands, Quebec, in the Gulf of St. Lawrence are younger.

- Geologists agree that the whole Permian period was one of environmental stress and change. The worst effects came at the end of the Permian period, 30 million years after the rocks here formed. The ensuing mass extinction killed off 90 per cent of all marine species and 70 per cent of land animals.

Also Nearby

Just 16 kilometres from Cavendish Beach is a historic building made of local sandstone, the Farmers' Bank of Rustico. Now a registered historic place, it was built by the local Acadian community in 1861–1863 to serve as a community bank.

To visit, follow Route 6 south and east about 13 kilometres to its intersection with Route 243 (Church Road, N46.41576 W63.29174) and turn left. Travel northeast about 1 kilometre to the Farmers' Bank of Rustico (N46.42321 W63.28337), which houses a museum.

For more information about the building, museum, and other attractions at the site, visit http://farmersbank.ca.

Farmers' Bank of Rustico.

The John Hamilton Grays had much to discuss, as depicted in this statue on Great George Street, part of Charlottetown's heritage walk.

Red and Brown

Historic Buildings of Local Permian Sandstone

An iconic statue in Charlottetown's historic district depicts Fathers of Confederation John Hamilton Gray of Prince Edward Island and John Hamilton Gray of New Brunswick deep in discussion. Presumably this confab would have taken place during the pivotal Charlottetown Conference of 1864, which laid the groundwork for Canadian nationhood.

While wrestling with political complexities, perhaps these two took time to discuss another pressing matter: red sandstone, or brown? In 1864 the region was entering a "golden age" for sandstone as an abundant, durable, and fire-resistant building material. The red variety is local to Prince Edward Island, while brown varieties underlie tracts of New Brunswick and neighbouring Nova Scotia.

Across Great George Street from the Grays' endless discourse stands St. Dunstan's Basilica, made of brown sandstone from French Fort Cove (site 41), New Brunswick. Province House is made of brown Nova Scotia sandstone. But within a few blocks numerous other stone buildings brighten the city with the colourful red, locally quarried variety, known simply as Island sandstone. More ornate buildings in the historic district combine the local red with imported brown stone.

Getting There

Driving Directions

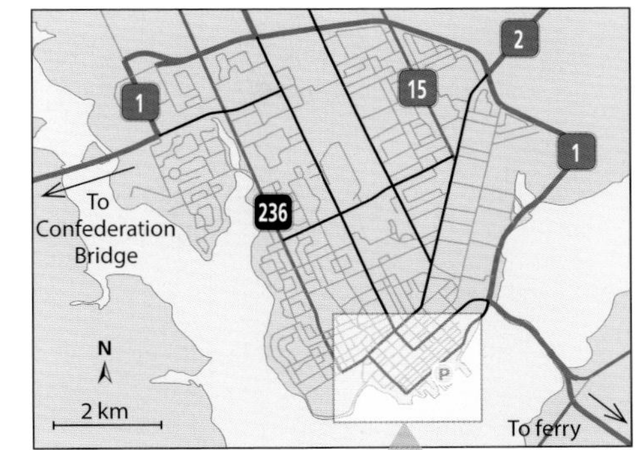

From the west, follow Route 1 toward Charlottetown, but go straight (N46.26152 W63.16928) at Capital Drive. Turn south (right) onto University Avenue and at Euston Street turn east (left). From Euston turn south (right) onto Prince Street. Continue past Water Street to the parking location.

From the east, follow Route 1 to Charlottetown and turn (N46.24134 W63.11385) onto Water Street. Continue on Water to Prince and turn south (left) for the parking location.

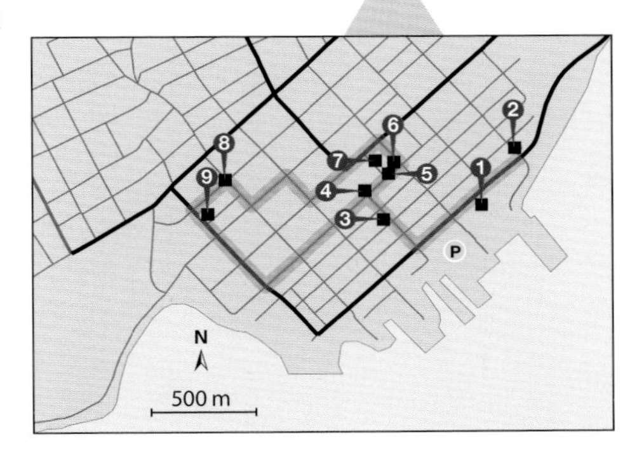

Where to Park

Parking Location: N46.23295 W63.12178

Park in the lot by Confederation Landing. Or, use another parking facility or on-street parking convenient to the walking tour.

Walking Directions

Many routes are possible. As an example, from the parking location follow Water Street east to the Brass House (1) and Railway Station (2). Backtrack on Water to Great George Street and walk north, passing St. Dunstan's Basilica (3) on the way to Province House (4). To view St. Paul's Rectory, Church Hall, and Church (5, 6, 7), walk east on Richmond to Prince, north on Prince to Grafton, and west on Grafton to Church. Farther west on Grafton, at Queen walk north, turning west onto Kent at City Hall. At Pownal walk north one block to view the Kirk of St. James (8). Turn west on Fitzroy to All Soul's Chapel (9). Return to the parking location via Rocheford, Richmond, and Great George.

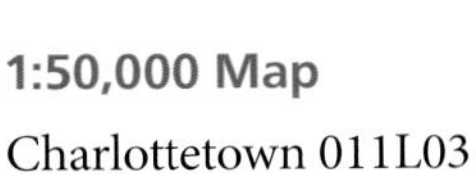

1:50,000 Map

Charlottetown 011L03

Provincial Scenic Route

Charlottetown

1. Brass House, 1876

Corner of Water and Hillsborough. Located in the city's former rail yard, this building is named for its past use as a workshop for fabricating and repairing brass components of trains. Built just a few years after the provincial railroad was established, its exterior walls are entirely Island sandstone, with well-preserved stonework details.

2. Railway Station, 1907

Corner of Water and Weymouth. This three-storey building replaced an older wooden station near the site. Not a station hotel, its upper floors provided office, meeting, and storage space for the railroad and its unions. Its exterior is red Island sandstone with brown Nova Scotia sandstone trim.

Stops 3 and 4 on this walking tour are historic buildings made of imported sandstone. St. Dunstan's Basilica (3) is made of stone from the Fish quarry in French Fort Cove, Miramichi (site 41). Province House (4) is made of sandstone from the Wallace quarry in Nova Scotia.

Rock Unit

Hillsborough River Formation, Pictou Group

Bedrock Map

Not available

5. St. Paul's Rectory, 1888

Corner of Richmond and Church. The rectory was designed and built by the same local team as the church itself. On its walls of homogeneous Island sandstone, chiselled, pecked, sawn, and hewn surfaces provide a variety of architectural details.

6. St. Paul's Church Hall, 1906

Corner of Richmond and Prince. This building is predominantly Island sandstone, with limited accents of Nova Scotia sandstone on the front wall.

7. St. Paul's Anglican Church, 1896

Corner of Grafton and Church. The church exterior is mainly Island sandstone, but with extensive trim of brown Nova Scotia sandstone. Local architect W.C. Harris designed the church, which was built by a Charlottetown firm, Lowe Brothers.

Rock Unit

Hillsborough River Formation, Pictou Group

Bedrock Map

Not available

8. Kirk of St. James, 1878

Corner of Fitzroy and Pownal. Local architect W.C. Harris was involved in the design of this Presbyterian church, but it was constructed by a firm from Nova Scotia. They used brown sandstone from the Wallace quarry, Nova Scotia. The doors, windows, and buttresses are accented in red Island sandstone.

9. All Soul's Chapel, 1888

Corner of Fitzroy and Rocheford. This tiny, gemlike chapel built of Island sandstone stands beside a larger brick church, St. Peter's Anglican Cathedral. Like St. Paul's Church and rectory (stops 5 and 6), it was designed by W.C. Harris and built by Lowe Brothers.

FYI

- Island sandstone used in the buildings of Charlottetown was likely quarried locally. For example, a known quarry is preserved as a national historic site (but not open to the public) on Mt. Edward Road.
- Most sandstone blocks in the buildings are homogeneous, but some are noticeably coarse grained, with scattered granules 1 or 2 millimetres across. Others have weathered to reveal cross-bedding.
- The Nova Scotian Wallace sandstone used in several of Charlottetown's historic buildings is part of a rock formation that continues into southern New Brunswick (see site 20, Cape Enrage, FYI).

Rock Unit

Hillsborough River Formation, Pictou Group

Bedrock Map

Not available

Also Nearby

St. Martin of Tours Roman Catholic Church (N46.16812 W63.16833) is on Route 19 in Cumberland (see map below), about 3.2 kilometres southwest of the turn-off for Skmaqn–Port-la-Joye–Fort Amherst National Historic Site of Canada.

The church was built in 1865–1868 of local Island sandstone. The bell tower, gable ends, and window surrounds are brick.

Eastern Prince Edward Island

If you are travelling in eastern Prince Edward Island, for example via Woods Island and the ferry from Nova Scotia, consider exploring other heritage buildings made of local stone, described below. All but the town hall were designed by W.C. Harris, who seems to have been partial to the red sandstone of his island home.

- St. Mary's Roman Catholic Church, corner of Longworth Street and Chapel Avenue (Route 305), Souris (N46.359297 W62.253066). Built in 1930 of Island sandstone with imported trim.
- Souris Town Hall, 75 Main Street (Route 2), Souris (N46.355073 W62.255032). Built in 1905 of Island sandstone with Wallace, Nova Scotia, sandstone trim.
- Kings County Courthouse, 60 Kent Street (Route 3), Georgetown (N46.184006 W62.533399). Built in 1887 of Island sandstone alternating with imported brown sandstone.
- St. Paul's Roman Catholic Church, 1133 Cambridge Road (Route 17A), Sturgeon (N46.104581 W62.533554). Built in 1873–1888 of Island sandstone with imported trim.

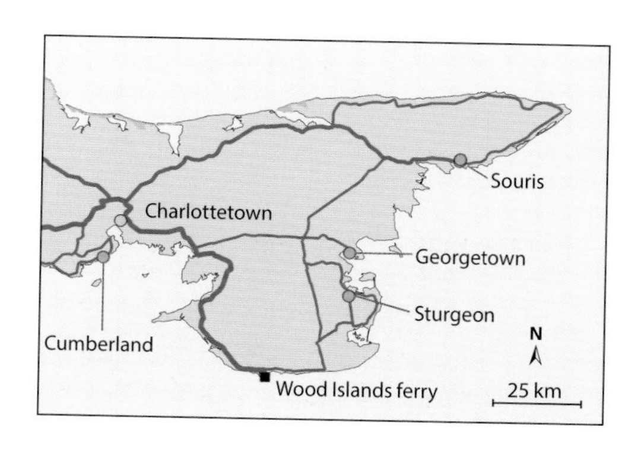

Exploring Further

Prince Edward Island Historic Places. http://www.gov.pe.ca/hpo/app.php (a searchable registry with details about the island's historic buildings).

This section provides geologic maps and rock descriptions covering the routes of three major highways in New Brunswick. Its aim is to help you appreciate the province's geological story as you pass by on your journey.

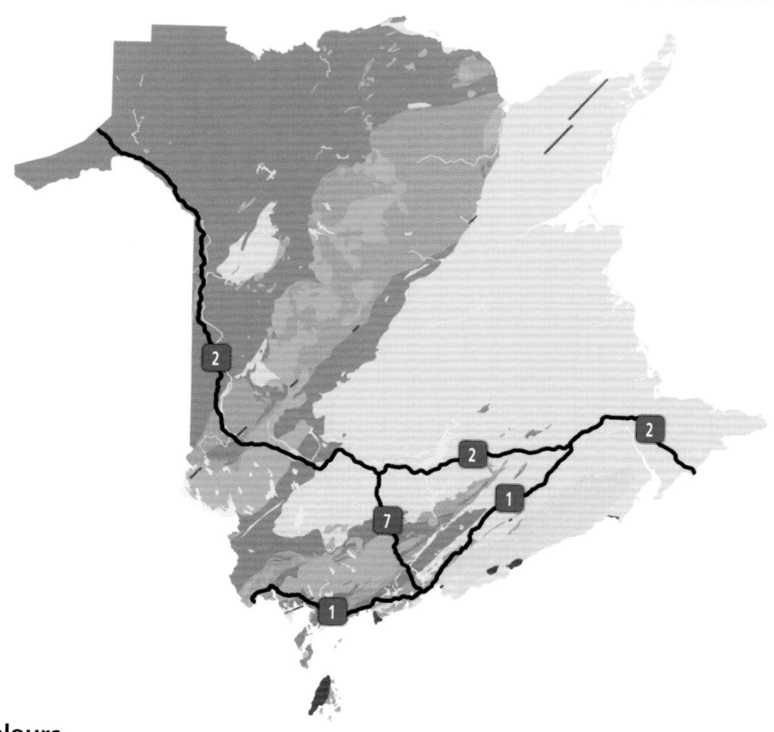

Map Colours

Colours on the seven detailed highway maps in this section correspond to the four geological stories outlined in the book's introduction.

A. Foundations **B. Collisions** **C. Pangaea** **D. Atlantic** See pages 18–26

Safety

Highway rock outcrops can be appreciated without compromising your safety or that of others. Please remember:

 Don't stop or exit your vehicle on highways except in an emergency.

 Don't slow down or take your eyes off the road to inspect highway rock exposures.

 Don't try to read while driving.

Highway Geology

Routes 1, 2, and 7

Three major highways cross the western and southern regions of New Brunswick. These well-engineered roadways expose a wide variety of rock formations, many of them not readily visible elsewhere.

Kilometre markers like this one on Route 2 near Florenceville-Bristol help travellers and road crews pinpoint highway locations.

Where Am I?

In New Brunswick, green-and-white kilometre markers have been placed at regular intervals along each major highway to help road crews and travellers identify highway locations. Exit numbers correspond to nearby marker numbers.

The markers always measure driving distances from the highway's westernmost point, so at a given location, markers in the eastbound and westbound lanes are identical. Travellers in the westbound lane will pass markers and exits in descending order.

The geologic maps for Routes 1, 2, and 7 show the location of markers every 10 kilometres, allowing you to estimate your position. The text describes locations in relation to markers or exits and also provides GPS waypoints. Either can be used to track your progress toward rocks of interest.

Virtual Tours

If you see rocks that make you curious while travelling New Brunswick's highways, don't risk trying to get a better look while driving. For safety's sake, consider a virtual tour of roadside rock formations using Google Maps Street View or a similar product—all the outcrops are visible there.

While travelling, use highway markers (photo, previous page) or exit numbers to remember interesting locations, or use the waypoints provided. Then "browse" the outcrop later online in the safety of a rest stop or travel accommodation, or at home.

Further Resources

Information about New Brunswick road conditions is available by phone at 511 or online at http://511.gnb.ca. Google Maps includes complete street-view coverage of New Brunswick highways. Visit www.google.ca/maps.

For detailed geologic maps of the routes, use the index map below to identify the National Topographic System (NTS) grid code of the area you are interested in (all the codes begin with 021, for example, 021N08). Visit the New Brunswick Department of Energy and Resource Development website at https://dnr-mrn.gnb.ca (search for "geology maps"). On the maps page, review the instructions and then click the corresponding grid number to access the PDF file.

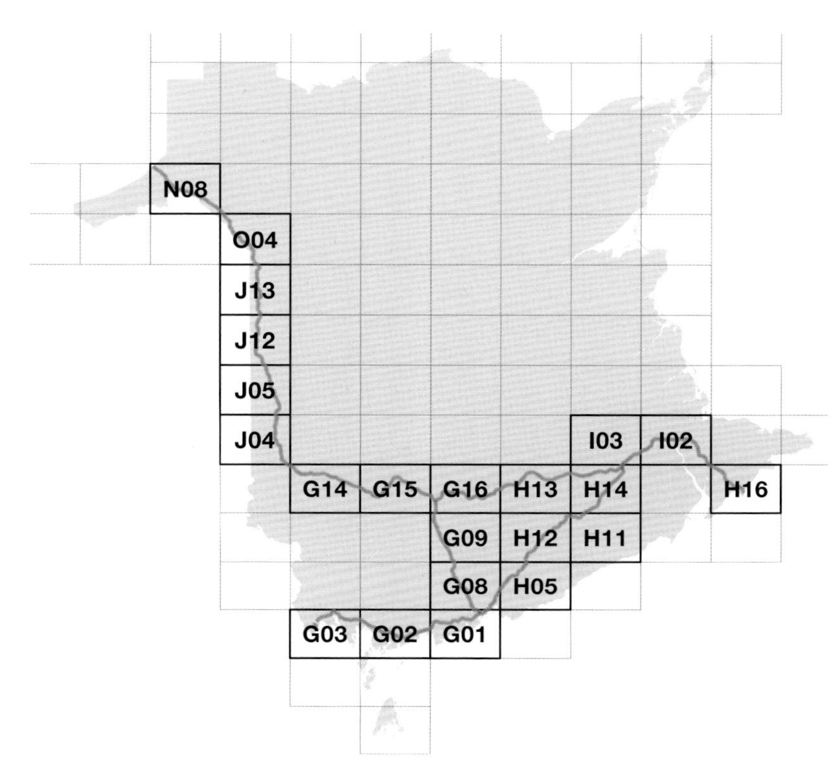

Route 1

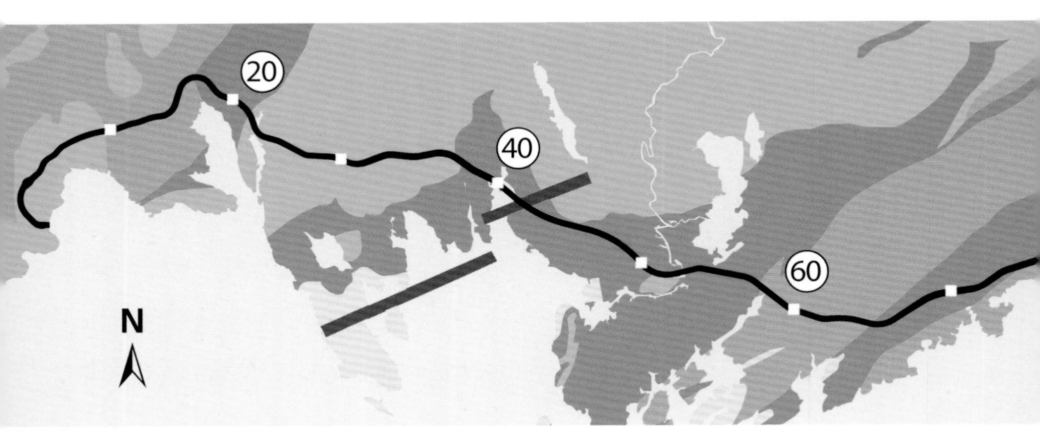

Markers 0 to 76, US Border to Pocologan. Route 1 begins at the US border near St. Stephen and broadly follows the Fundy coast eastward to Pocologan. Along the way it crosses a complex tectonic collage including four ancient terranes (parts of Ganderia and Avalonia), a volcanic arc (Kingston belt), and its back-arc basin (Mascarene basin).

0–10 St. Stephen Gabbro

This Silurian intrusion has some pale grey granitic areas as well as more abundant exposures (some of them rust stained) of dark grey gabbro.

- Between the border station and marker 1 (N45.16204 W67.31284), an area of pale grey granitic rock can be seen beside the eastbound lane.
- Farther along, past marker 3 at the start of the exit lane for Exit 4 (N45.17900 W67.32353), good examples of the dark, rusty gabbroic rocks of the intrusion are on both sides of the highway.

See related rocks: sites 1, Todds Point; 4, St. George.

10–18 St. Croix Terrane

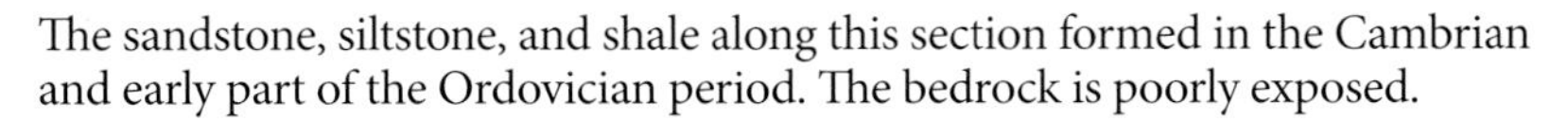

The sandstone, siltstone, and shale along this section formed in the Cambrian and early part of the Ordovician period. The bedrock is poorly exposed.

- Between markers 17 and 18, east of Route 755 (N45.24158 W67.18744), two quarries straddle the highway and expose the deep-ocean sediments typical of this terrane. The contact with overlying rocks of the Mascarene basin is exposed in the quarry but is not visible from the highway.

18–24 Mascarene Basin

The Silurian Mascarene back-arc basin contains carbon-rich sedimentary rocks (dark grey; some with rusty stains) as well as volcanic ash layers (pale pink, grey, or dark green) and gabbro dykes (black).

- East of marker 20 is a truck weigh station, and east of that (N45.22015 W67.15029), before marker 22, is a high outcrop of folded, faulted, dark grey sedimentary and igneous layers.

See related rocks: sites 5, Greens Point; 6, Herring Cove.

24–37 Bocabec Complex

Between markers 24 and 25, the Waweig River lies near the boundary between rocks of the Mascarene basin and those of the Bocabec complex. This well-exposed Silurian intrusion has both granitic (pale pink or pale grey) and gabbroic (dark grey) areas.

- Between markers 29 and 30 (N45.19737 W67.08663) is a good example of the dark grey gabbroic rocks of the complex.
- Immediately east of marker 34 (N45.19764 W67.03086) beside the eastbound lane is a contrasting example of the distinctly pink-tinged granitic rocks.
- The gabbro outcrops near Bocabec (between markers 36 and 37, for example, N45.19805 W66.99650 by the westbound lane) are especially significant: less than 1 kilometre south of the highway is the site from which grave markers of similar rock were quarried for victims of the *Titanic* disaster.

See related rocks: sites 1, Todds Point, FYI; 4, St. George.

37–57 Mascarene Basin

Between marker 37 and Exit 39, Route 1 enters another part of the Mascarene basin. This section exposes numerous outcrops of volcanic rock. (The dyke mapped as crossing Route 1 is not visible from the road.)

- On either side of marker 40 (N45.18300 W66.96361) on the westbound side is a series of high outcrops of greenish grey sedimentary rock and basalt.
- Between markers 48 and 49 (for example, N45.14826 W66.86916) is a series of brick-red outcrops of Mascarene rhyolite.

See related rocks: sites 5, Greens Point; 6, Herring Cove

57–65 New River Terrane

Between markers 57 and 58, the Letang River marks a major tectonic boundary between the Silurian Mascarene basin and the Ediacaran New River terrane.

- One of the most dramatic outcrops on Route 1 lies between markers 58 and 59 (N45.12107 W66.75662) on the westbound side of the highway. Sheets and fingers of 550-million-year-old orange-pink granitic rock mingle with contrasting zones of dark greenish grey gabbro.
- A less conspicuous but significant outcrop of the 620-million-year-old Blacks Harbour grey granite can be seen beside the eastbound lane between markers 63 and 64 (N45.10765 W66.69860).

See related rocks: site 7, Pea Point, Also Nearby.

65–71 Kingston Belt

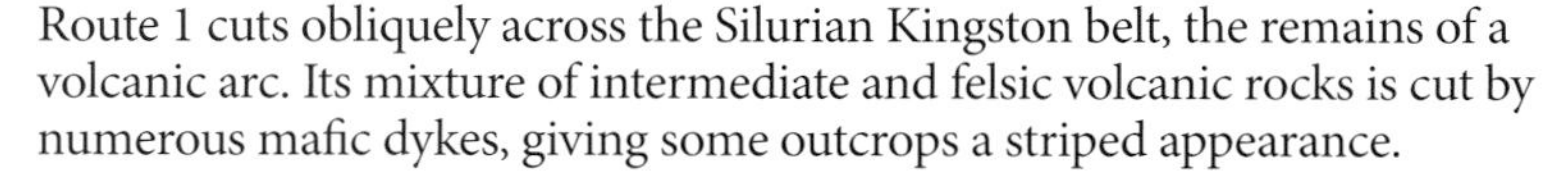

Route 1 cuts obliquely across the Silurian Kingston belt, the remains of a volcanic arc. Its mixture of intermediate and felsic volcanic rocks is cut by numerous mafic dykes, giving some outcrops a striped appearance.

- The resistant igneous rocks of the Kingston belt have provided many long outcrops along this section of the highway. Watch for a good example of the complex interlayering of felsic and mafic components of the belt between Exit 69 and marker 70 (N45.12144 W66.62801).

72–76 Kingston Belt (Pocologan Metamorphic Suite)

The Pocologan River marks the approximate boundary between the Kingston belt and its high-pressure counterpart, the Pocologan metamorphic suite. Strengthened by metamorphism and extreme deformation during the Carboniferous period, these mylonitic rocks form dramatic outcrops. In certain light conditions the aligned minerals glisten and reveal their strong linear fabric.

- Between markers 74 and 75 is a long series of high outcrops in the form of steeply tilted, smooth-sided grey slabs. Outcrops beside the westbound lane (for example, N45.13732 W66.56848) are most dramatic.

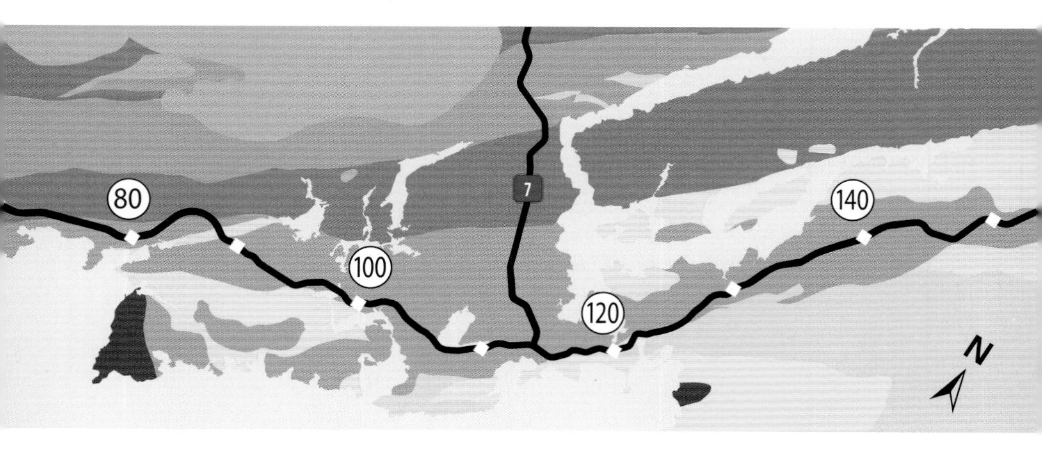

Markers 76 to 154, Pocologan to Quispamsis. Still roughly following the Fundy shore, Route 1 continues east from Pocologan past Saint John, then turns inland, skirting north around the Caledonia Highlands. This section of the highway is dominated by rocks of the ancient Brookville terrane.

76–83 Brookville Terrane

This part of the Brookville terrane is deformed in parallel with the adjacent Pocologan metamorphic suite, but the terrane's mainly granitic and dioritic intrusions gradually become less deformed as you travel east. The outcrops along this section of the highway are small and highly fractured.

83–86 Kingston Belt (Pocologan Metamorphic Suite)

As the highway swings north, crossing the Lepreau River between markers 82 and 83, it makes a brief excursion back into the Kingston belt's Pocologan metamorphic suite.

- Between marker 84 and Exit 86 (N45.18716 W66.46737), high, nearly vertical slabs of grey mylonite stand beside the westbound lane.

86–119 Brookville Terrane

East of Exit 86 the highway turns southward and re-enters the Brookville terrane. Through marker 111, the rocks along this section are mainly felsic intrusions (dark red to pale grey depending on their composition). At Exit 112 the highway crosses into Ediacaran marble and quartzite of the Green Head Group.

- Just before and continuing past marker 101 beside the westbound lane (N45.19676 W66.29073) is a tall, pale grey outcrop of the Prince of Wales granite, approximately 550 million years old.
- Near Exit 114 (N45.23035 W66.14125), bright grey areas of marble beside the westbound lane are eye-catching.

See related rocks: site 15, Rockwood Park.

The rock exposures along Route 1 and its access ramps in Saint John are extensive and dramatic, but the traffic is typically busy. Please avoid temptation and keep your eyes on the road while driving in and around the city.

119–127 Caledonia Terrane

Between markers 118 and 119, just after Route 1 passes under Lancaster Avenue, it crosses from the Brookville terrane into the northern edge of the Caledonia terrane. Here the Caledonia terrane includes Ediacaran volcanic rocks as well as Cambrian sedimentary rocks.

See related rocks: sites 14, King Square West; 15, Rockwood Park, FYI; 18, Fundy Trail Parkway; 19, Point Wolfe.

127–154 Brookville Terrane

Between markers 126 and 127 the highway passes from the Caledonia terrane back into the Brookville terrane and weaves through areas of granitic intrusions, gneiss, marble, and quartzite, all from the Ediacaran period.

Please keep your eyes on the road while driving in and around Saint John.

- East of Saint John, between markers 144 and 145 just east of Elliot Road (N45.42887 W65.90970), a long series of marble outcrops includes bright light grey and mottled darker grey varieties.
- Immediately west of marker 152 (N45.46547 W65.84472) beside the westbound lane is a long outcrop of pink and orange granitic rock that intruded the marble and other metasedimentary rocks of the Brookville terrane.

See related rocks: site 15, Rockwood Park.

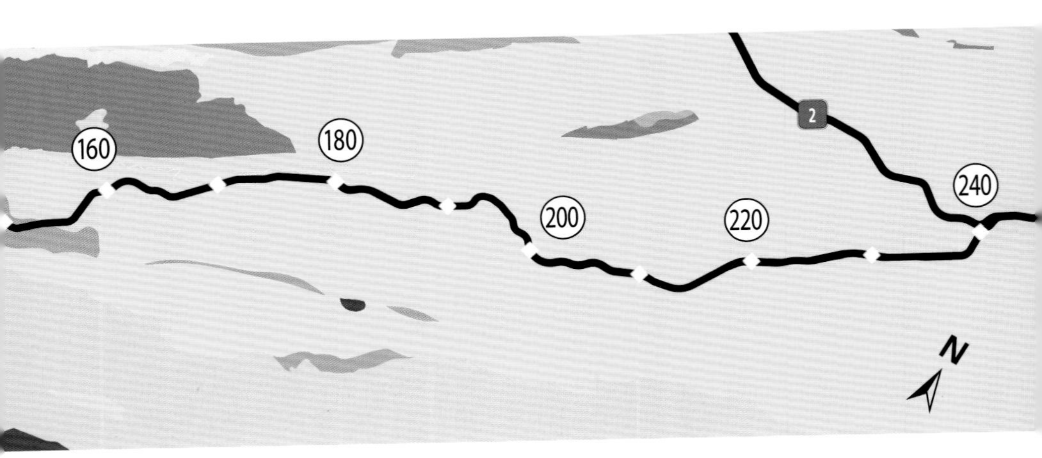

Markers 155 to 240, Quispamsis to Route 2. From Quispamsis Route 1 follows the Kennebecasis River northeast; it ends at Route 2 near Salisbury. This entire section of the highway passes over rocks of the Maritimes basin.

155–240 Maritimes Basin

In general, the exposed sedimentary layers along Route 1 become younger from west to east. See p. 25 for a list of the basin's principal rock groups.

155–190 Horton Group

Numerous outcrops of deformed, layered sedimentary rock appear along this section. Layers rich in sand resist weathering and form prominent beds; shale layers rich in mud weather more easily and shed piles of crumbly debris that may obscure parts of the outcrop. Numerous pebbles make conglomerate layers look lumpy and mottled.

- Between Exit 166 and marker 168 (N45.57455 W65.74905) beside the eastbound lane is a succession of conspicuous layers of light brown sandstone tilting gently to the south.
- Between markers 182 and 183 (N45.66904 W65.62557) beside the eastbound lane is a long section of steeply tilted, light brown sandstone and dark grey shale.
- About 200 metres east of marker 188 (N45.69792 W65.56997) in the centre median are outcrops of albertite, an oily shale of the kind used to make the first kerosene.

See related rocks: sites 2, St. Andrews; 7, Pea Point; 22, Hillsborough, Also Nearby.

190–200 Sussex and Windsor Groups

Around Sussex, Route 1 crosses faulted slices of the Sussex and Windsor groups. Weakened by deformation and weathering, they are poorly exposed along the highway. Weathering of these rocks has produced fertile soil and gentle topography, making the area well suited for agriculture.

See related rocks: site 22, Hillsborough.

200–220 Mabou and Cumberland Groups

Few outcrops are visible along this stretch of mainly flat-lying sandstone.

- Immediately west of marker 203 (N45.73721, W65.42135) on both sides of the eastbound lane are long, low outcrops of dark red sandstone.

See related rocks: sites 11, Lepreau Falls; 20, Cape Enrage; 21, Hopewell Rocks.

220–240 Pictou Group

Sandstone of the Pictou Group typically forms a gently rolling landscape, so there was little need to cut through rock when building the highway. No notable outcrops occur along this stretch of Route 1.

See related rocks: sites 23, Minto; 24, Fredericton Junction; 39, Cap-Bateau; 40, Grande-Anse to Neguac; 41, French Fort Cove.

Route 2 (Trans-Canada Highway)

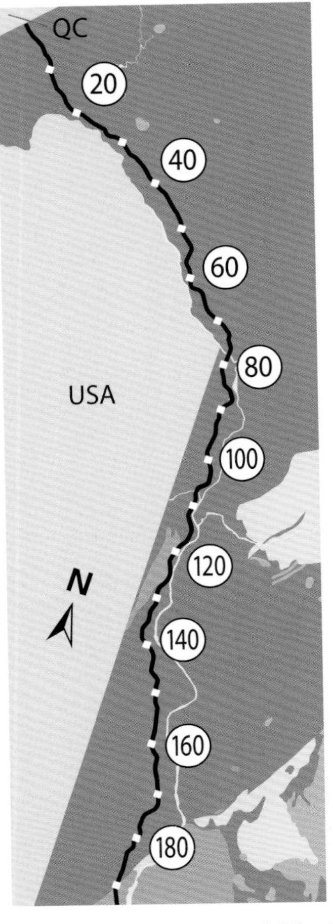

Markers 0 to 190, Quebec border to Woodstock. Route 2 begins at the Quebec border in northwestern New Brunswick and roughly follows the St. John River southward to Woodstock. This entire section of the highway crosses sedimentary rocks of the Matapédia basin.

0–190 Matapédia Basin

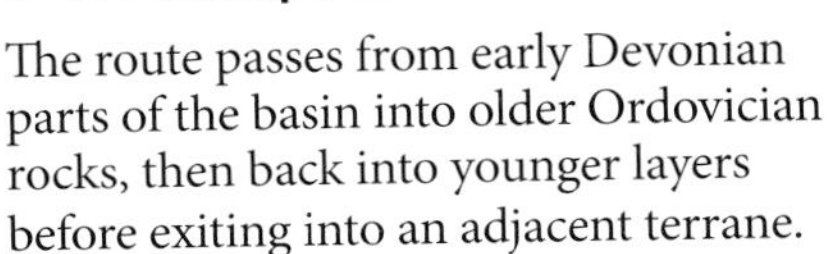

The route passes from early Devonian parts of the basin into older Ordovician rocks, then back into younger layers before exiting into an adjacent terrane.

0–38 Devonian slate

During the Devonian period, mud accumulated in a deep marine basin, then was compressed, forming crumbly slate. A major fault around marker 38 separates the Devonian slate from similar but older rocks to the south.

- Between Exits 13A and 13B on the eastbound side (N47.40150 W68.36357) is a typical outcrop, broken into large, straight-edged blocks with fine, near vertical cleavage.
- Similar outcrops occur in the westbound lane between markers 24 and 25 (for example, N47.36038 W68.23507).

See related rocks: site 31, Edmundston.

38–190 Ordovician and Silurian shale and limestone

These Ordovician and Silurian deepwater deposits are mainly shale, limestone, and mixtures of the two, intruded by mafic dykes and then deformed by later events.

- On the high ground east of Exit 51 (N47.22221 W67.94640) several complex folds are traced by light limestone layers among the darker shale on the westbound side (but with a good view from the eastbound lane).
- At marker 105 (N46.83733 W67.71849) both lanes cut through high outcrops of complexly deformed layers, including some dark, greenish gabbro dykes.
- Beside marker 147 (N46.49793 W67.68133) are younger, dark shales relatively poor in limestone.
- Limestone-rich rocks similar to those farther north recur between markers 158 and 160 (for example, N46.39146 W67.64055).

Rock exposures of the Matapédia basin end just south of Exit 188 to the US and I-95.

See related rocks: site 30, Grand Falls.

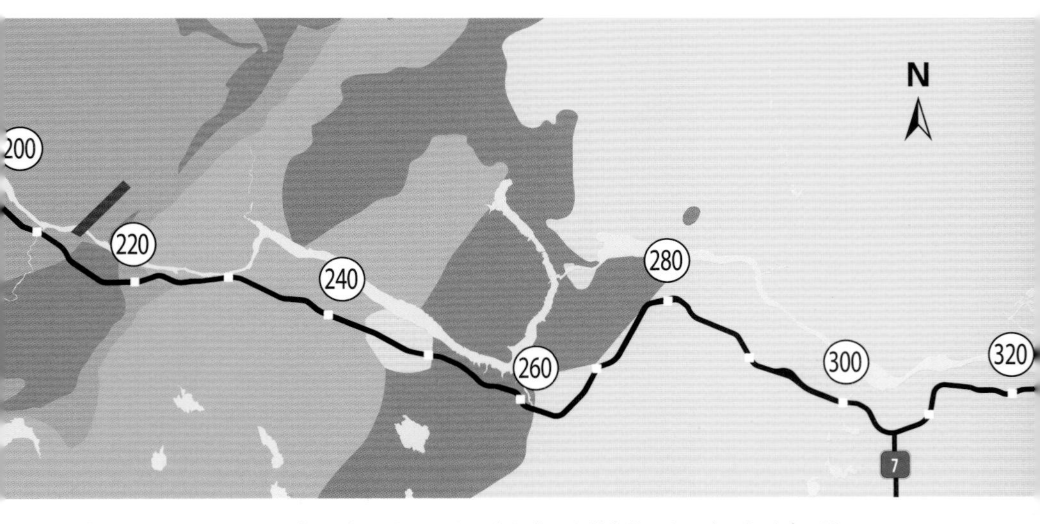

Markers 190 to 324, Woodstock to Swan Creek Lake. Still following the St. John River, Route 2 leads south and east from Woodstock, past the intersection with Route 7 to Swan Creek Lake. The highway crosses a dizzying array of geologic settings in rapid succession—the rocks of an ancient terrane, three different basins, and a complex felsic intrusion—before entering the Maritimes basin.

190–212 Miramichi Terrane

These rocks are the oldest exposed along Route 2. Most are metamorphosed sand and silt layers (now quartzite and slate) originally deposited off the shores of Ganderia just before it separated from Gondwana. Near Meductic are younger but related rocks of the terrane, including felsic volcanic and iron-rich sedimentary rocks.

- In the eastbound lane, watch for outcrops as the highway climbs the hill between markers 195 and 196 (N46.08563 W67.57080).
- In the westbound lane, they are better seen near the crest of the same hill, just past marker 196 (N46.08147 W67.56918).

See related rocks: site 29, Hays Falls.

212–218 Matapédia Basin

Here rocks of the Miramichi terrane along Route 2 are interrupted by outcrops of overlying Silurian and Devonian rocks of the Matapédia basin. These younger layers include siltstone, shale, and basalt.

- Between markers 213 and 215 (for example, N45.96576 W67.45707) both lanes pass by long outcrops of steeply tilted siltstone.
- Between markers 216 and 218 (for example, N45.95413 W67.43334, eastbound lane only) the basalt looks darker and more blocky than the siltstone.

218–221 Miramichi Terrane

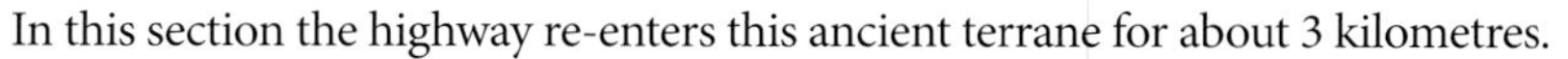

In this section the highway re-enters this ancient terrane for about 3 kilometres.

- Near the terrane's eastern edge, long outcrops line both lanes near marker 220 (N45.94911 W67.38282).

221–244 Pokiok Batholith

Several different kinds of granitic rock occur within the early Devonian Pokiok batholith. In many outcrops a process called exfoliation has broken these rocks into horizontal layers that, at highway speeds, might be mistaken for sedimentary bedding. But the sharp, sometimes curved, broken edges and commonly massive size of the blocks signals their hard, igneous character.

- Between markers 222 and 223 (N45.95473 W67.35392) both lanes pass beside long exposures of light grey, sharp-edged, blocky outcrops of grey tonalite.
- About 16 kilometres farther east, on the high ground between markers 238 and 239 (N45.92461 W67.16118) high walls of pink granite line the roadway.

See related rocks: site 28, McAdam.

244–250 Maritimes Basin

For about 5 or 6 kilometres, perched across a boundary between much older rocks, is a little pod of Carboniferous sandstone.

- Walls of sandstone beside both lanes on the high ground around marker 246 (for example, N45.89810 W67.07606) are typical of this rock formation.

See related rocks: sites 23, Minto; 24, Fredericton Junction.

250–261 Fredericton Trough

For a little highway drama, along this stretch watch for a series of features marking the faulted contact between younger, flat-lying rocks of the Maritimes basin and the older, deformed rocks of the Fredericton trough.

- Between markers 250 and 251 is a valley marking the faulted western boundary of the Fredericton trough.
- West of the valley are flat-lying layers formed from Carboniferous river deposits (N45.88561 W67.02851). But just 500 metres to the east are rocks of the Fredericton trough (N45.88496 W67.02079), formed in deep seawater about 100 million years earlier during the tumultuous Silurian period.
- A dramatic fold in these older rocks by marker 261 (best seen on the north side in both east- and westbound lanes; N45.84005 W66.90780) attests to their complex history.

See related rocks: site 26, Mactaquac Dam.

261–324 Maritimes Basin

Route 2 passes onto rocks of the Maritimes basin around marker 261, and from there to the Nova Scotia border remains in this geological setting. Due to past warping and faulting in the basin, the age of rocks exposed along the highway varies.

261–263 Mabou Group

A single exposure of gently tilted, red sandstone provides a glimpse of this narrow band of rocks formed during the middle of the Carboniferous period.

- About 500 metres east of marker 261 (N45.83840 W66.90067) is a conspicuously reddish sandstone. Its broken edges have been softened by weathering.

See related rocks: sites 11, Lepreau Falls; 21, Hopewell Rocks.

263–324 Pictou Group

This rock formation, deposited late in the Carboniferous period, typically appears along Route 2 as low ledges of brown, flat-lying sandstone.

- Both lanes pass through a long corridor of sandstone beginning just past marker 264 (for example, N45.84192 W66.85432).
- You may be puzzled to notice some highly deformed, grey rocks near Hanwell (eastbound lane, immediately west of Deerwood Drive: N45.93011 W66.75996; westbound lane, west of marker 278: N45.92822 W66.76494). There, the highway skims along a fault boundary, exposing much older rocks of the Fredericton trough.
- Near marker 324 (N45.85441 W66.27060) on either side of Swan Creek Lake, some of the layers look rough because they are pebble conglomerates.

See related rocks: sites 23, Minto; 24, Fredericton Junction.

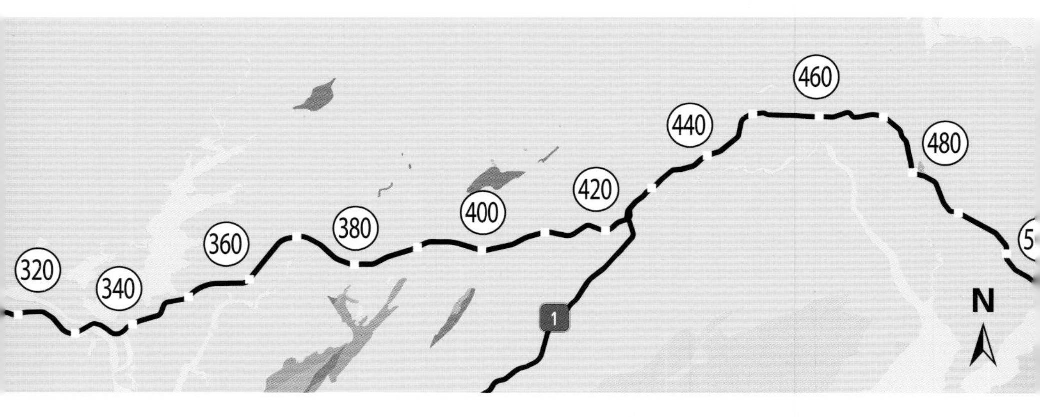

Markers 324 to 515, Swan Creek Lake to the Nova Scotia border. From Swan Creek Lake, Route 2 wends along the valleys of the St. John River, Grand Lake, and the Canaan River, then crosses the Canaan River. The route continues east around Moncton, then turns south, approximately following the course of the Petitcodiac River. Along the whole section from Swan Creek Lake to the Nova Scotia border, Route 2 crosses rocks of the Maritimes basin.

324–515 Maritimes Basin

This section of Route 2 is located entirely within the Maritimes basin. It crosses alternating areas of younger and older Carboniferous sedimentary rocks.

324–396 Pictou Group

Like the adjacent area to the west, this part of the highway is underlain by relatively young rocks of the Maritimes basin's Pictou Group.

- East of marker 325 (N45.84888 W66.25883), outcrops of brown sandstone beside the eastbound lane are similar to those spanning Swan Creek Lake.

396–440 Older rocks of the Maritimes basin

Even geologic maps do not all agree on the age of the rock layers along this part of the highway due to their limited exposure, but in general they include older parts of the Maritimes basin—the Horton, Sussex, and Mabou groups.

- Between markers 400 and 401 immediately east of the Route 880 overpass (N45.94138 W65.41721), look for tilted layers of reddish sandstone and conglomerate.
- Between markers 430 and 431 (N46.034011 W65.082435), best seen in the westbound lane, is a series of long, low walls of dark reddish brown sandstone.

440–480 Cumberland and Pictou groups

Around Moncton, the rocks underlying Route 2 formed late in the Carboniferous period as parts of the Cumberland and Pictou groups. They are not exposed along the highway.

480–490 Horton and Mabou groups; Devonian granite

In this section, Route 2 again passes through an area where most of the underlying rock formed early in the Carboniferous period (Horton and Mabou groups). A small Devonian granite intrusion provides some variety (and a useful source of aggregate) near Memramcook.

- Travelling eastbound between markers 481 and 482, you may catch a glimpse of the Memramcook aggregate quarry from the eastbound lane (N46.04688 W64.57292); the westbound lane passes close by quarry operations (N46.04624 W64.56678).
- At marker 488 a long wall of grey sandstone (N46.00156 W64.51783) appears beside the westbound lane and may be visible eastbound.

490–515 Cumberland and Pictou groups

This section is underlain by sandstones formed late in the Carboniferous period (Cumberland and Pictou groups). No noticeable exposures of rock can be seen along the highway.

Route 7

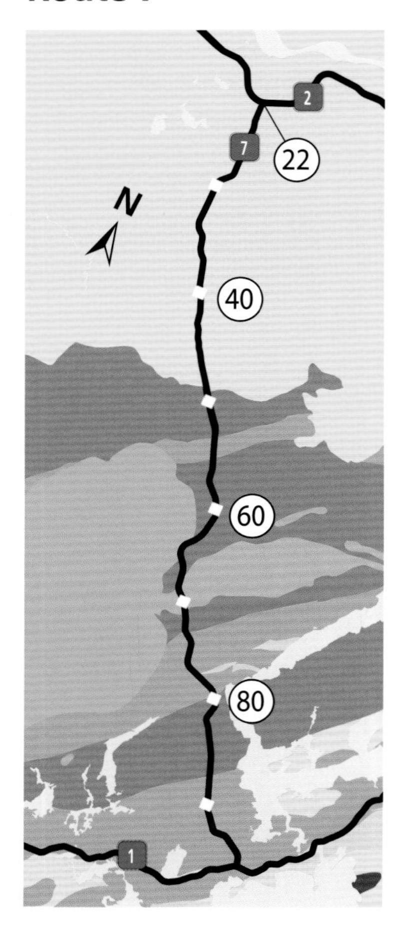

Markers 22 to 97, Oromocto to Saint John. Route 7 diverges south from Route 2 at Oromocto and continues south, eventually ending at Route 1 just west of Saint John. It first crosses increasingly older rocks of the Maritimes basin before cutting through part of the Fredericton trough. Between Welsford and Saint John the highway crosses a complex region assembled by tectonic collisions and major faults, including two ancient Ganderian terranes, a volcanic arc and back-arc basin, and large granitic intrusions.

22–49 Maritimes Basin

From Route 2 to about marker 45, the highway crosses a sequence of flat-lying sandstone layers about 315 million years old. They aren't well exposed, but you may catch a glimpse of some low, brown or reddish brown outcrops along the way.

- As you drive south, older rocks are exposed, including 360-million-year-old, pale, tilted layers of volcanic rock on the high ground between markers 47 and 48 (N45.59386 W66.40300).

See related rocks: site 27, Harvey.

49–53 Fredericton Trough

Here, sandstone as well as finer-grained siltstone and shale layers are tilted, in some places quite steeply. The turbidites of the Fredericton trough contain impurities like iron, sulphur, and organic carbon that create multicoloured outcrops in some locations.

- The long wall of tilted layers immediately north of marker 50 (southbound lane only, N45.57645 W66.38742) is typical of rocks formed in this deep marine basin.

See related rocks: site 26, Mactaquac Dam.

53–55 Annidale Terrane

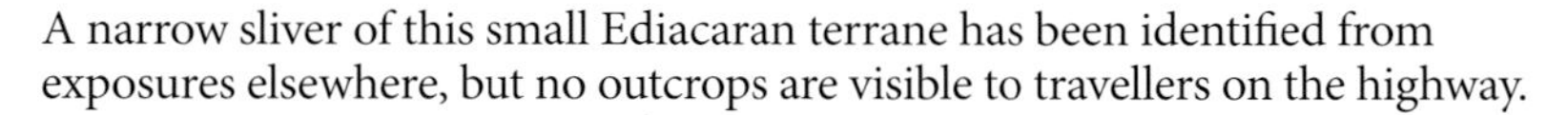

A narrow sliver of this small Ediacaran terrane has been identified from exposures elsewhere, but no outcrops are visible to travellers on the highway.

55–73 Mascarene Basin and Younger Granitic Intrusions

This Silurian back-arc basin contains carbon-rich sedimentary rocks (dark; some with rusty stains) as well as volcanic ash layers (pale pink, grey, or dark green) and gabbro dykes (black). The highway also takes two brief excursions into younger granitic rocks that intruded the basin. To see these, watch for tall, blocky outcrops in shades of pink, salmon, or brick red.

- One of the first large outcrops, just north of marker 58 (N45.51204 W66.34567), displays angular, pale grey volcanic rocks.
- Between markers 63 and 65, the 360-million-year-old Mount Douglas granite forms some dramatic outcrops, for example, between markers 64 and 65, just north of where Route 7 crosses Route 101 (N45.46055 W66.34610).
- Back in the Mascarene basin, between markers 66 and 67 (N45.44372 W66.33928) and again beside marker 68 (N45.43000 W66.33509) are dark, steeply tilted sedimentary layers cut by wavy fingers of granite related to one of the nearby plutons.
- The 420-million-year-old Welsford granite makes an appearance about halfway between markers 69 and 70 (N45.41990 W66.32660) with a high wall of pale pink.
- Near the basin's southern boundary at Exit 71 (N45.40651, W66.31413) high, steeply tilted layers lie near a major fault and appear sheared and fractured.

See related rocks: sites 1, Todds Point; 4, St. George; 5, Greens Point; 6, Herring Cove.

73–80 New River Terrane

About 200 metres north of marker 73, Route 7 crosses a small inconspicuous stream which marks the trace of a major fault separating the Mascarene basin from the New River terrane. This stretch of the highway provides numerous large outcrops of old (Ediacaran) rocks, including resistant granitic and metamorphosed sedimentary and volcanic rock.

- About 200 metres south of marker 73 (N45.39050 W66.30110) is a 500-metre-long outcrop of metamorphosed sedimentary rock.
- Between markers 78 and 79 (N45.35847 W66.25641) is a long low outcrop of granitic rock known to be about 550 million years old.

80–87 Kingston Belt

B

Exit 80 approximately marks the boundary between the New River terrane and the Silurian Kingston belt. The Kingston belt's volcanic arc produced a complex mixture of volcanic ash layers and related shallow granitic intrusions (both commonly pinkish), cut by numerous metamorphosed gabbro dykes (greenish with rusty stains). In long outcrops you may be able to notice the colour variations.

- Extending from markers 81 to 83 is a chain of representative outcrops (for example, N45.32816 W66.24025).
- The rocks exposed at either end of the Exit 86 interchange (N45.30733 W66.22392) provide further examples of Kingston belt igneous activity.

87–97 Brookville Terrane

A

Just south of marker 87, Route 7 dips to cross a stream valley along the trace of a large fault between the Kingston belt and the more ancient Brookville terrane. The highway passes through areas of granite, marble, and metamorphosed siltstone, a combination that extends all the way into Saint John.

- Between markers 87 and 88 (N45.29050 W66.21550) an outcrop of Brookville granite is cut by a mafic dyke (best seen beside the northbound lane).
- From markers 91 to 93 (for example, N45.25905 W66.19692), the rocks take on a very different appearance—dark grey or dark brown, in many places mottled with bright white. This is a mixture of siltstone and marble.

See related rocks: sites 10, Barnaby Head; 15, Rockwood Park.

Glossary

allochthonous Part of a rock sequence or formation that has been tectonically displaced onto an adjacent terrain and thus did not form where it is now found. ah-LOCK-thon-us

alluvial fan A wedge-shaped accumulation of mostly sand and gravel deposited where a stream flows from a mountain valley onto an adjacent lowland.

amphibole A silicate mineral rich in iron and magnesium, typically black or dark green and typically occurring as prismatic crystals in igneous or metamorphic rocks. It is the dominant mineral in amphibolite, a rock typically formed by metamorphism of mafic intrusions. AM-fih-bowl, am-FIB-o-lite

andesite A volcanic rock with a composition intermediate between those of basalt and rhyolite. AN-de-zite or AN-de-site

anaerobic Characterized by a lack of oxygen. an-ah-ROE-bik

anhydrite A calcium sulphate mineral, $CaSO_4$, formed by precipitation from sea water or by dehydration of a related mineral, gypsum. an-HIGH-drite

aplite A type of granite having an even, fine-grained texture and very few or no dark minerals. AP-light

arkose A type of sandstone in which at least a quarter of the sand is feldspar rather than quartz. AR-koze

ash flow An avalanche-like surge of hot volcanic gases and vapourized magma produced by an explosive volcanic eruption; or, the products of such an event.

Avalonia A fragment separated from the continent Gondwana during the Cambrian or Ordovician period and now recognized as a number of terranes within the Appalachian orogen. Named for the Avalon Peninsula of Newfoundland, where it was first recognized. av-ah-LONE-ee-ah

basalt An extrusive igneous rock formed primarily from plagioclase and pyroxene, similar in composition to gabbro but with mainly fine-grained crystals too small to see with the naked eye. bah-SALT

batholith A complex igneous intrusion at least 100 square kilometres in area. Typically granitic in composition, batholiths may include several distinct but related rock types formed by separate pulses of molten rock. BATH-o-lith

basin A low region of the Earth's surface surrounded by relatively high elevations. This includes ocean basins but also basins formed by warping or faulting of continental crust.

bimodal In igneous rocks, characterized by two distinct, contrasting rock types (for example, granite and gabbro) rather than by a single rock type or a range of related types. BY-MODE-ul

biotite A dark brown or black mica that is high in iron and magnesium.

back-arc basin A low-lying area beside an island arc, formed by rifting of the arc and subsequent sea-floor spreading along the rift. The pulling-apart motions that form the basin are paradoxically caused by subduction of an older ocean plate on the other side of the arc.

black shale A fine-grained, typically thinly layered sediment that is rich in organic matter and contains sulphur-rich minerals, formed in oxygen-poor conditions, often in a basin with limited water circulation.

Bouma sequence A five-part sequence of sedimentary layers named for its discoverer, Arnold Bouma, and characterized by diminishing grain size (coarsest layers first, finest last) and a defined series of rock textures. Some parts of the sequence may be missing, but they are never out of order; each sequence records the passage of a single turbidity current. BOOM-ah sequence

calcite A calcium carbonate mineral, $CaCO_3$, commonly found filling cracks and other voids in rocks, or making up carbonate rocks such as limestone (sedimentary) or marble (metamorphic).

caldera A large crater-like depression created by the collapse of a magma chamber after it is emptied by a catastrophic volcanic eruption. call-DARE-ah

carbonate A rock consisting of any combination of calcium and magnesium carbonate, including limestone and dolostone (sedimentary) and marble (metamorphic).

case hardening A process that occurs during the loss of water from a porous rock, in which chemical elements carried by the pore water are deposited near the rock surface as the water evaporates, creating a weather-resistant outer layer.

chlorite A greenish silicate mineral with a scaly structure, somewhat similar to mica but easily scratched, commonly found in rocks affected by low-grade metamorphism. KLOR-ite

clast A fragment of any size produced by the physical disintegration of rock material and subsequently incorporated into a sedimentary rock.

cleavage In rocks, a fabric caused by deformation and metamorphism in which the alignment of flat mineral grains makes the rock weak and easily fractured along any plane parallel to this feature.

columnar jointing A network of cracks formed during the cooling of magma and resulting in long, parallel, typically hexagonal prisms (columns) of rock. The orientation of the resulting columns may vary depending on how cooling progressed.

conglomerate A sedimentary rock containing a high proportion of rock fragments larger than sand (pebbles, cobbles, or boulders), typically cemented in a finer matrix.

continental shelf The nearshore portion of the continental margin, that is, the shallow marine environment between the shore and the continental slope.

cross-bedding Inclined deposits of sediment within a larger horizontal bed, formed along the edge of a ripple, bar, dune, or similar feature.

crystal tuff A rock formed by consolidation of volcanic ash containing visible crystals or crystal fragments, typically of quartz and/or feldspar. The crystals formed in a cooling magma chamber beneath the volcano prior to eruption. KRIS-tul TOOF

dacite A volcanic rock of composition intermediate between andesite and rhyolite, composed of quartz, plagioclase, and minor amounts of dark minerals such as hornblende or biotite. DAY-site

delta A typically wedge-shaped accumulation of sediment at the mouth of a river, deposited as water flow slows down, causing sediment to settle.

detrital minerals Mineral grains surviving intact from the disintegration of pre-existing rock, then transported, redeposited, and incorporated into a sedimentary rock. dih-TRY-tul

diorite An intrusive igneous rock typically containing two parts plagioclase to one part hornblende and/or pyroxene, and little or no quartz, that is, with a composition intermediate between those of granite and gabbro. DYE-oh-rite

dyke A narrow, tabular or sheet-like discordant igneous intrusion formed when molten rock flows into a crack in an existing rock.

epiclastic In the context of volcanic processes, formed by the reworking, transport, and redeposition of erupted material by any surface process including landslides, flowing water, or ice. epp-ih-KLASS-tik

epidote A silicate mineral containing calcium, aluminum, and iron, formed during low-grade metamorphism of rocks or by the alteration of igneous minerals such as pyroxene or amphibole. EPP-ih-dote

exfoliation A complex geologic process, typically affecting igneous rocks of uniform texture, that causes outer layers of nearly uniform thickness to separate from the underlying mass. The process occurs on both large and small scales, leading to rounded shapes whether of entire landforms or individual boulders.

fabric In rocks, the pattern visible due to the spatial arrangement and average shape of the mineral grains, rock fragments, or other constituents.

fault A plane or narrow zone along which a rock mass fractures and displacement occurs. A fault's intersection with the Earth's surface is known as a fault trace, in some cases visible as a landform such as a fault scarp (a steep, linear slope between high and low ground).

feeder Part of the underground network of cracks, channels, and conduits through which magma travels toward the surface in a volcanically active region.

feldspar A group of silicate minerals in which silica, aluminum, and varying amounts of sodium, calcium, and potassium combine in a framework-like crystal lattice. Varieties include plagioclase and potassium feldspar.

felsic A rock composition characterized by a predominance of the light-coloured minerals quartz and feldspar.

ferric oxide, ferrous oxide Two forms of iron oxide formed in contrasting conditions. Orange ferric oxide, Fe_2O_3, forms in the presence of plentiful oxygen, while grey ferrous oxide, FeO, forms in oxygen-poor environments.

flood basalt Mafic magma characterized by high volume and low viscosity, which pours out of long fissures onto the surface of the Earth, covering large areas rather than building up distinct volcanic landforms.

flute cast A long, narrow welt on the bottom of a sandstone bed, typically with one end well defined and bulbous and the other end merging into a surrounding flat area; literally the cast (mould) of a channel-like depression scoured into the top of the underlying layer, typically by a turbidity current.

foreland basin A depressed region of continental crust, warped downward by an adjacent mass thrust onto the continent, for example, by collision or obduction.

gabbro An intrusive igneous rock typically containing equal portions of plagioclase and pyroxene. Similar in composition to basalt but with larger crystals easily visible to the naked eye. GAB-roe

Ganderia A fragment separated from the continent Gondwana during the Ordovician period and now recognized as a number of terranes within the Appalachian orogen. Named for the area around Gander, Newfoundland, where it was first recognized. gan-DEER-ee-ah

gneiss A coarse-grained, foliated rock with alternating bands of dark and light minerals. Gneiss can form from either sedimentary or igneous rocks under conditions of high-grade metamorphism. NICE

Gondwana A supercontinent of the geologic past that included areas found in the present-day continents of the southern hemisphere: India, Africa, South America, Australia, and Antarctica. gond-WAH-nah

graded bedding Sedimentary layering in which the size of the particles gradually decreases upward within a single deposit or bed. Grading occurs because the settling rate of a particle in water is proportional to its size (the smaller the particle, the slower the rate).

granite An intrusive igneous rock typically containing equal parts of quartz, plagioclase, and potassium feldspar, with or without small amounts of mica and/or amphibole. GRAN-it

granodiorite A felsic intrusive rock similar to granite but with less potassium feldspar. gran-uh-DYE-oh-rite

greywacke A marine sedimentary rock, typically grey in colour, characterized by a mixture of angular quartz and feldspar grains, tiny rock fragments, and clay. Sometimes simply known as wacke. grey-WACK-ee

gypsum A calcium sulphate mineral, $CaSO_4 \cdot 2H_2O$, formed by precipitation from sea water or by hydration of a related mineral, anhydrite. JIP-sum

half-graben A structure formed in rocks under tension, in which one side of a crustal block slips down along a fault, rotating and tilting the block. Typically the movement forms an asymmetrical valley having a steep side along the fault and a gradual slope on the other side. HAFF GRAH-ben

hornblende A variety of amphibole.

hyaloclastic Characterized by volcanic rock fragments formed when hot lava erupted underwater, forming a thin, glassy crust that then shattered due to thermal shock. HI-ah-low-KLASS-tik

Iapetus Ocean An ocean of the geologic past that formed as the supercontinent Rodinia split apart, separating Laurentia from other continental areas. ee-APP-ah-tus

immature Of sedimentary rock, characterized by angular clasts in a wide range of sizes and by materials that are typically broken down during prolonged exposure to surface conditions (rock fragments, dark minerals, quartz, feldspar, clay).

inclusion A rock fragment occurring within an igneous rock, having been carried along while the igneous rock was in a molten state and typically originating in rock units through which the magma has travelled.

inlier A defined area of old rock that is surrounded by unrelated younger rock units, typically due to differences in erosion rate, history of uplift, and/or the effects of folding.

intrusion A body of molten rock that has travelled upward through the Earth's crust and invaded or displaced pre-existing rock; or, the process by which this occurs.

island arc A curved chain of islands formed by volcanic activity above a subduction zone.

laminations Millimetre-scale layering in sedimentary rock formed by the deposition of small amounts of material during a succession of minor changes in water conditions or sediment characteristics.

Laurentia A continent of the geologic past (Proterozoic eon) that included areas found in present-day North America and parts of Europe. lore-REN-chee-ah

lithic tuff A rock formed by consolidation of volcanic ash and containing rock fragments. The fragments are typically igneous rock torn away from the volcano by the explosive eruption of the ash. LITH-ik TOOF

lithosphere The outermost, rigid layer of Earth material, about 100 kilometres thick, consisting of both oceanic and continental crust and the upper mantle, and comprising the Earth's tectonic plates. LITH-oh-SFEER

magma Molten rock.

magma chamber An underground reservoir of molten rock.

mantle The most voluminous layer of the Earth's interior, lying between the iron-rich outer core and the silica-rich crust.

matrix The fine-grained part of a rock in which larger crystals or rock fragments are embedded.

Meguma The largest of several fragments separated from the continent Gondwana during the Paleozoic era and now recognized as a terrane within the Appalachian orogen. The word is derived from the Mi'kmaq name for their own people. meh-GUE-mah

metamorphic core complex An area of high-grade metamorphic rocks, typically in a dome-like shape bounded by shear zones, that was exhumed (moved rapidly upward) in response to extension of the continental lithosphere.

metamorphism The process by which the minerals in a rock recrystallize in response to changing conditions of temperature and pressure. met-ah-MOR-fizz-em

mica Any one of a group of aluminum-rich silicate minerals having crystals that form stacks or "books" of shiny, thin layers (see biotite, muscovite). MY-cah

microcontinent A fragment of continental crust rifted from a larger, pre-existing continental mass.

mudstone A sedimentary rock that originated as very fine silt and clay and that lacks the easily parted layering of shale.

muscovite A type of mica lacking iron or magnesium, recognized by its transparent or silvery colour. MUSK-ah-vite

magma mingling A process in which two volumes of molten rock (typically one felsic and the other mafic) intrude into the same space and interact while cooling, but without significant blending. For magmas of strongly contrasting composition, differences in crystallization temperature and viscosity render them immiscible, like oil and water, allowing them to remain distinct while interacting.

meander Of a river, to alter course by processes of erosion and deposition, forming a series of bends that widen and migrate downstream over time. mee-ANN-der

mylonite A fine-grained rock type characterized by strong linear and/or planar fabric formed during metamorphism and intense deformation in a fault zone. MY-lah-nite

nuée ardente Literally, "glowing cloud"; a type of volcanic eruption characterized by fast-moving, turbulent masses of hot volcanic gas, ash, and rock and mineral debris that flow down the mountainside. NOO-ay ar-DAHNT

oceanic ridge The long, narrow, mountainous area of an ocean basin along which new ocean crust is formed by a continuous process of rifting and volcanic activity.

ophiolite A sequence of rock types formed at ocean ridges, including most or all of: deep ocean sediments, pillow basalts, sheeted dykes, gabbroic intrusions, and mantle peridotite. OH-fee-ah-lite

orbicular granite A felsic igneous rock characterized by rounded, concentric structures (orbs) of varying mineral composition. In some varieties, the entire orb is igneous in composition; in others, metamorphosed sedimentary rock lies at the orb centre. or-BIK-you-lar GRAN-it

orogeny, orogen The process of mountain building, including metamorphism, deformation, and igneous activity. An orogen is the result of this process as preserved in the rock record, regardless of whether the region is still mountainous or has been worn down by erosion. oh-RODGE-enn-ee, ORE-oh-jenn

Pangaea A supercontinent of the geologic past (approximately 300 to 175 million years ago). pan-GEE-ah

passive margin A continental margin situated adjacent to old, stable oceanic crust not subject to significant tectonic activity; not a tectonic plate boundary.

pegmatite An igneous rock type characterized by large (centimetre-scale or greater) mineral grains, typically occurring in veins or pods within an igneous intrusion and in some cases hosting rare minerals containing boron, lithium, or fluorine. PEG-mah-tite

peperite A rock formed by the interaction of magma and wet or lightly consolidated sediment, with a variety of outcomes including complex mingling or inclusions of one material in the other. PEPP-er-ite

pillow lava A form of basalt exhibiting bulbous shapes formed as lava erupts under water. Some pillows shatter during this process, forming pillow breccia.

plagioclase A type of feldspar containing calcium and/or sodium rather than potassium. PLAJ-ee-o-klaze

plumose fracture A pattern on the fractured surface of a rock that widens from a point of origin and is marked by a series radiating lines and concentric arcs, reminiscent of a large, feathery plume; a series of radiating as the fracture propagates.

porphyry An igneous rock characterized by conspicuous crystals (known as phenocrysts) in a fine-grained matrix, resulting from a two-stage cooling process of the original magma: slow cooling during which the phenocrysts form, followed by rapid cooling to form the fine-grained matrix. POUR-for-ee

potassium feldspar A type of feldspar containing potassium rather than calcium and/or sodium. Potassium feldspar is typically pinkish due to iron oxide coatings and lends granite its characteristic colour.

pumice A glassy felsic volcanic rock with a frothy, open structure formed by the eruption into the air of a bubbly, gas-rich magma that hardens as it falls to earth.

pyrite A metallic mineral, iron sulphide, also known as "fool's gold" due to its shiny, yellow crystals. PIE-rite

pyroclastic Describing volcanic eruptions dominated by the violent ejection of hot ash, droplets of magma, rock fragments, and other material. pie-roe-CLASS-tick

pyroxene A silicate mineral rich in iron and magnesium, similar to amphibole but with a simpler crystal structure that forms at higher temperatures. PEER-ox-een

Rheic Ocean An ocean of the Paleozoic era that formed between Gondwana and several microcontinents, including Ganderia, Avalonia, and Meguma. REE-ick or RAY-ick

rhyolite A volcanic rock equivalent in composition to granite. RYE-o-lite

rift A regional-scale break in the Earth's crust, caused by tectonic forces causing tension within oceanic or continental crust, typically resulting in deep, steep-sided valleys and, in some cases, volcanic activity.

rillenkarren A feature that forms on outcrops of water-soluble rock types such as limestone or gypsum, caused over decades of time by rainwater slowly dissolving grooved paths down steep surfaces. RILL-en-KAR-en

roche moutonnée A landform in which exposed bedrock has been glacially sculpted into an asymmetric shape, with a shallower slope on one side (upstream with respect to ice flow) and a steeper slope on the other (downstream) side; an equivalent English term is sheepback. ROSH MOO-ton-ay

Rodinia A supercontinent of the geologic past (approximately 1,000 to 750 million years ago). roe-DIN-ee-ah

shale A sedimentary rock type formed from muddy sediment and having a flaky texture due to the alignment of clay minerals.

shear zone Similar to a fault in that both are caused by relative movement of adjacent blocks of crust, but in a shear zone, the movement is "smeared out" across a wide band instead of being focused along a single plane.

siltstone A sedimentary rock formed from silt, that is, sediment with a grain size between that of sand and mud.

slump fold A fold formed not by regional tectonic pressures but rather by the sliding and crumpling of partially consolidated sedimentary layers, for example down an inclined basin margin, possibly triggered by an earthquake or simply by material instability.

sole mark A general term for any of a variety of structures located on the base of a sedimentary layer, whether formed by currents, the weight of dense overlying sediment, or other cause.

spherulite A nodule-like, roughly spherical cluster of radially arranged minerals common in glassy volcanic material, especially of felsic composition. SFEER-you-lite

striation In glaciated landscapes, one of a set of usually parallel gouges or scratches on the surface of a rock, caused by glacial abrasion as rock-laden ice scrapes across a rocky landscape. stry-AY-shun

strike-slip fault A fault along which opposing crustal blocks move sideways past one another, also known as a transcurrent fault.

subduction A process by which one tectonic plate moves beneath another along their common boundary and sinks into the Earth's mantle. The setting where this occurs is known as a subduction zone. sub-DUCK-shun

subvolcanic Characteristic of a process that took place or a rock that formed near, but under, the surface of the Earth as part of a volcanic episode. Subvolcanic rocks may superficially resemble volcanic rocks, requiring microscopic examination to make the distinction.

supercontinent A landmass that includes all or many regions of the Earth's continental crust, assembled through a series of continental collisions.

tectonic Pertaining to the large-scale structural features of the Earth's crust, and to the forces and processes that deform the crust. tek-TAH-nik

temnospondyl Any member of a large group of four-legged animals generally considered to be the primitive ancestors of present-day amphibians. They first appeared early in the Carboniferous period and were wiped out in the Triassic-Jurassic extinction event. tem-no-SPON-dul

terrane A fault-bounded fragment broken from one tectonic plate and later joined to another during plate collision, recognized as such by its distinctive rock units and separate sedimentary, igneous, and/or metamorphic history as compared to adjacent parts of an orogen.

texture A characteristic of rocks that describes the size, shape, and arrangement of their constituent parts.

tonalite An igneous rock containing primarily quartz and plagioclase, with lesser amounts of dark minerals such as biotite and amphibole. TONE-ah-lite

trace fossil Preserved evidence of animal activity, for example, in the form of tracks, burrows, feeding marks, or resting places.

trachyte Volcanic rock composed primarily of potassium feldspar with small amounts of quartz, plagioclase feldspar, and dark minerals such as biotite or pyroxene. Intrusive rock with the same mineral composition is known as syenite. TRACK-ite

transcurrent fault A fault along which opposing crustal blocks move sideways past one another, also known as a strike-slip fault.

tuff A rock formed by consolidation of volcanic ash deposited either on land or into a body of water from which it later settled. TOOF

turbidite A sedimentary rock deposited in deep ocean water by turbulent, avalanche-like currents of liquefied sediment (known as turbidity currents) flowing down the continental slope. TUR-bah-dite

unconformity An erosion surface preserved in a sequence of rock layers, representing a period of time during which no sediment was deposited, or if deposited, was subsequently removed by erosion. In an angular unconformity, sedimentary layers below the unconformity are tilted with respect to the younger layers, signifying an intervening cycle of deformation, uplift, and subsidence. un-con-FOR-mit-tee

vent A crack or pipe-like opening through which lava, steam, or other volcanic emissions reach the Earth's surface.

vesicular Containing empty cavities formed when bubbles are trapped in lava as it cools and hardens. veh-SICK-you-lar or vee-ZICK-you-lar

volcanic arc A linear or slightly curved pattern of volcanic activity above a subduction zone. A volcanic arc located entirely within oceanic crust is called an island arc.

volcanic ash A collective term used to refer to fine-grained fragments (less than 2 millimetres wide) of volcanic glass, crystals, and rock ejected during an explosive eruption.

volcanogenic massive sulphide (VMS) deposit A type of mineral deposit dominated by sulphides of copper and zinc, formed on the ocean floor by the circulation of hot, mineral-rich fluids along mid-ocean ridges and other ocean-floor rifts.

zircon A zirconium silicate mineral, $ZrSiO_4$, commonly formed as tiny crystals in felsic igneous rocks and useful for measurement of geologic time due to their uranium content.

Index of Place Names

Prince Edward Island

Image Credits

All photographs, maps, and diagrams are by Martha Hickman Hild and Sandra M. Barr, except as noted below:

Front cover, photo of Hopewell Rocks, New Brunswick © Kevin Snair, http://creativeimagery.ca. Published with permission.

Page 47, map of regional dyke locations adapted from McHone & colleagues, 2014, *Atlantic Geology*, vol. 50, p. 69, fig. 2.

Page 81, photo of Badwater alluvial fan, Death Valley, California © Marli Miller, http://geologypics.com. Published with permission.

Page 81, sketch of temnospondyl track republished with permission of Canadian Science Publishing, from Sarjeant & Stringer, 1978, *Canadian Journal of Earth Sciences*, vol. 25, p. 600, fig. 5A; permission conveyed through Copyright Clearance Center, Inc.

Page 93, map of Partridge Island block adapted from Park and colleagues, 2014, *Canadian Journal of Earth Sciences*, vol. 51, p. 3, fig. 2.

Page 107, map of Avalonia in New England and Atlantic Canada adapted from Hibbard & colleagues, 2006, *Geological Survey of Canada "A" Series Map 2096A*, sheet 2.

Page 111, map of Triassic basins adapted from Hamblin, 2004, *Geological Survey of Canada Open File 4678*, p. 27, fig. 2.

Page 117, photo of Walton Glen gorge by Sue Johnson. Published with permission.

Page 138, photo of Maid Marion dragline from the Minto mural by Ron Sajack. Published with permission of the mural's owner, Richard Veenhuis.

Page 138, photos of fossil plants taken with assistance from the Public Library, Village of Minto.

Page 138, Village of Minto emblem graphic provided by the Village of Minto. Published with permission.

Page 143, photo of Amazon River, image no. OLI-2014194 courtesy of the Earth Science and Remote Sensing Unit, NASA Johnson Space Center, http://eol.jsc.nasa.gov.

Page 155, caldera diagram adapted from Williams, 1951, *Scientific American*, vol. 185, No. 5, p. 51.

Page 163, detrital zircon age diagram adapted from Fyffe & colleagues, 2009, *Atlantic Geology*, vol. 45, p. 124, fig. 8f.

Page 167, map of Matapédia basin adapted from Wilson, 2017, New Brunswick Dept. of Energy and Resource Development, Geological Surveys Branch, Memoir 4, p. 2, fig. 2.

Page 177, photo of Mount Carleton summit by Reginald Wilson. Published with permission.

Page 203, map of Bathurst Mining Camp compiled from Thomas & colleagues, *Geological Survey of Canada Open File D3887*, p. 1, fig. 1.

Page 206, photo of sea arch by Michael Parkhill. Published with permission.

Page 221, map of Prince Edward Island geology adapted from van de Poll, 1989, *Atlantic Geology*, vol. 25, p. 24, fig. 1.

Geological maps on pages 18, 20, 22, 24, 26, 34, 51, 67, 77, 115, 121, 147, 155, 159, 163, 177, 187, 203, and all highway geology (pages 233–248) were prepared using data obtained from New Brunswick Department of Energy and Resource Development's 1:500,000 bedrock geology dataset (www.snb.ca/geonb1/e/DC/catalogue-E.asp) and related resources (https://www2.gnb.ca/content/gnb/en/departments/erd/energy/content/minerals.html) under the GeoNB Open Data Licence (http://www.snb.ca/e/2000/data-E.html). Our use of these data in no way suggests any official endorsement of the maps or of this book. Under the terms of the licence, no representation or warranty of any kind is made with respect to the accuracy or completeness of the information. Please consult the licence terms for additional details.

All road maps were produced using datasets obtained online from Canada's Geospatial Data Extraction resource (http://maps.canada.ca/czs/index-en.html) under the Open Government Licence—Canada (open.canada.ca/en/open-government-licence-canada). Although every effort has been made to provide helpful directions based on the datasets, under the terms of the licence, no representation or warranty of any kind is made with respect to the accuracy or completeness of the information. Please consult the licence terms for additional details.

About the Authors

Sandra (left) and Martha (right) on the Maliseet Trail near Hays Falls, New Brunswick.

Dr. Martha Hickman Hild received her PhD in Earth Sciences in 1976 from the University of Leeds, UK. Early in her career, she co-directed a research laboratory and lectured in geology. Subsequently she worked as an editor in both technical and educational publishing and as an award-winning news researcher. She is now a freelance writer and editor. The writing style and innovative format she developed for her first book, *Geology of Newfoundland*, led to its selection as "Best Guidebook 2014" by the Geosciences Information Society.

Dr. Sandra M. Barr received her PhD in Geology in 1973 from the University of British Columbia. Since 1976, she has been teaching geology at Acadia University in Wolfville, NS, where she is a Professor in the Department of Earth and Environmental Science. Widely recognized for her field-based studies of Appalachian geology, she has authored and co-authored hundreds of scientific publications and produced numerous reports and maps, many focused on the rocks of Maritime Canada. She is co-editor of the journal *Atlantic Geology* and from 2006 to 2019 served as book editor for the Geological Association of Canada.